Heat and Thermodynamics

Heat and Thermodynamics

Hilary D. Brewster

Oxford Book Company
Jaipur ▪ New Delhi

Published By

Oxford Book Company

Regd. Off. "Bony Residency", Gate No. 2
Opp. Tilak Public School, Vishweswariya Nagar,
Gopalpura Road, Jaipur - 302018, India
Ph: 0141 - 2761280, Telefax: 2761381
email - oxfordbookjpr@gmail.com
website - oxfordbookcompany.com

Delhi Office: 102, 1st Floor, Satyam House,
4327/3, Ansari Road, Darya Ganj,
New Delhi - 110002, India
Ph: 011-45652440

Distributed By

KEYA INTERNATIONAL PRESS LLP
E-mail: keyainternationalpressllp@gmail.com

Published in India, 2025

ISBN : 9789380179087

Price: ₹ 1895.00

Typeset by Shivangi Computers, Jaipur

Printed and bound in India by Thomson Press Ltd.,
New Delhi

Contents

Preface

Heat is basic science that deals with energy and has long been an essential part of Engineering Curricula all over the world. It was developed during the eighteenth and nineteenth century during a time when Temperature and Heat were not well understood yet. Its development was driven by the need for an improved theoretical understanding pSteam Engines invented at the same time. It evolved as a rather formal and elegant theory that proved to be of great importance to Engineers.

Thermodynamics is the science of Energy, *Heat, Work, Entropy and Spontaneity of processes*. It is closely related to Statistical Mechanics from which many Thermodynamic relationships can be derived. While dealing with processes in which systems exchange Matter or Energy, Classical Thermodynamics is not concerned with the rate at which such processes take place, termed Kinetics.

The book contains detailed descriptions of *Modern Techniques in Thermodynamics*. The aim of the book is to make the subject matter broadly accessible to advanced students, whilst at the same time providing a reference text for graduate scholars and research scientists active in the field.

Hilary D. Brewster

1

Introduction

HEAT

In physics, heat, symbolized by *Q*, is energy transferred from one body or system to another due to a difference in temperature. In thermodynamics, the quantity *TdS* is used as a representative measure of heat, which is the absolute temperature of an object multiplied by the differential quantity of a system's entropy measured at the boundary of the object. Heat can flow spontaneously from an object with a high temperature to an object with a lower temperature.

The transfer of heat from one object to another object with an equal or higher temperature can happen only with the aid of a heat pump. High temperature bodies, which often result in high rates of heat transfer, can be created by chemical reactions (such as burning), nuclear reactions (such as fusion taking place inside the Sun), electromagnetic dissipation (as in electric stoves), or mechanical dissipation (such as friction).

Heat can be transferred between objects by radiation, conduction and convection. Temperature is used as a measure of the internal energy or enthalpy, that is the level of elementary motion giving rise to heat transfer. Heat can only be transferred between objects, or areas within an object, with different temperatures (as given by the zeroth law of thermodynamics), and then, in the absence of work, only in the direction of the colder body (as per the second law of thermodynamics). The temperature and phase of a substance subject to heat transfer are determined by latent heat and heat capacity. A related term is thermal energy, loosely defined as the energy of a body that increases with its temperature. The first law of thermodynamics states that the energy of a closed system is conserved. Therefore, to change the energy of a system, energy must be transferred to or from the system. Heat and work are the only two mechanisms by which energy can be transferred to or from a control mass. Heat is the transfer of energy caused by the temperature difference. The unit for the amount of energy transferred by heat in the International System of Units SI is the joule (J), though the British Thermal Unit and the calorie are still occasionally used in the United States. The unit for the rate of heat transfer is the watt (W = J/s)

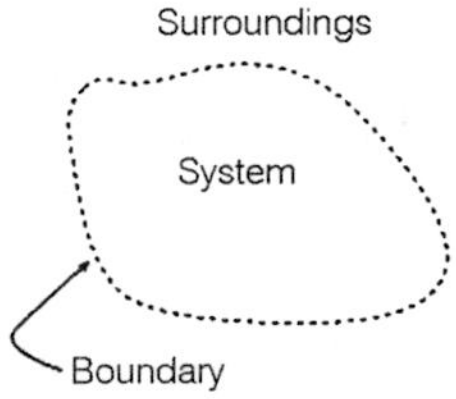

Heat Q can flow across the boundary of the system and thus change its internal energy *U*. Heat transfer is a path function (process quantity), as opposed to a point function (state quantity). Heat flows between systems that are not in thermal equilibrium with each other; it spontaneously flows from the areas of high temperature to areas of low temperature. When two bodies of different temperature come into thermal contact, they will exchange internal energy until their temperatures are equalized; that is, until they reach thermal equilibrium.

The adjective *hot* is used as a relative term to compare the object's temperature to that of the surroundings (or that of the person using the term). The term heat is used to describe the flow of energy. In the absence of work interactions, the heat that is transferred to an object ends up getting stored in the object in the form of internal energy

Specific heat is defined as the amount of energy that has to be transferred to or from one unit of mass or mole of a substance to change its temperature by one degree. Specific heat is a property, which means that it depends on the substance under consideration and its state as specified by its properties. Fuels, when burned, release much of the energy in the chemical bonds of their molecules. Upon changing from one phase to another, a pure substance releases or absorbs heat without its temperature changing. The amount of heat transfer during a phase change is known as latent heat and depends primarily on the substance and its state.

THERMAL ENERGY

Thermal energy is a term often confused with that of heat. Loosely speaking, when heat is added to a thermodynamic system its thermal energy increases and when heat is withdrawn its thermal energy decreases.

In this point of view, objects that are hot are referred to as being in possession of a large amount of thermal energy, whereas cold objects possess little thermal energy. Thermal energy then is often mistakenly defined as being synonym for the word heat. This, however, is not the case: an object cannot possess heat, but only energy.

The term "thermal energy" when used in conversation is often not used in a strictly correct sense, but is more likely to be only used as a descriptive word. In physics and thermodynamics, the words "heat", "internal energy", "work", "enthalpy" (heat content), "entropy", "external forces", etc., which can be defined exactly, i.e. without recourse to internal atomic motions and vibrations, tend to be preferred and used more often than the term "thermal energy", which is difficult to define.

NOTATION

The total amount of energy transferred through heat transfer is conventionally abbreviated as Q. The conventional sign convention is that when a body releases heat into its surroundings, $Q < 0$ (-); when a body absorbs heat from its surroundings, $Q > 0$ (+). Heat transfer rate, or heat flow per unit time, is denoted by:

$$\dot{Q} = \frac{dQ}{dt}.$$

It is measured in watts. Heat flux is defined as rate of heat transfer per unit cross-sectional area, and is denoted q, resulting in units of watts per square metre, though slightly different notation conventions can be used.

ENTROPY

In 1854, German physicist Rudolf Clausius defined the *second fundamental theorem* (the second law of thermodynamics) in the mechanical theory of heat (thermodynamics): "if two transformations which, without necessitating any other permanent change, can mutually replace one another, be called equivalent, then the generations of the quantity of heat Q from work at the temperature T, has the *equivalence-value*:"

$$\frac{Q}{T}$$

In 1865, he came to define this ratio as entropy symbolized by S, such that, for a closed, stationary system:

$$\Delta S = \frac{Q}{T}$$

and thus, by reduction, quantities of heat äQ (an inexact differential) are defined as quantities of TdS (an exact differential):

$$\delta Q = TdS$$

In other words, the entropy function S facilitates the quantification and measurement of heat flow through a thermodynamic boundary

DEFINITIONS

In modern terms, heat is concisely defined as energy in transit. Scottish physicist James Clerk Maxwell, in his 1871 classic *Theory of Heat*, was one of the first to enunciate a modern definition of "heat". In short, Maxwell outlined four stipulations on the definition of heat. One, it is "something which may be transferred from one body to another", as per the second law of thermodynamics.

Two, it can be spoken of as a "measurable quantity", and thus treated mathematically like other measurable quantities. Three, it "can *not* be treated as a substance"; for it may be transformed into something which is not a substance, e.g. mechanical work.

Lastly, it is "one of the forms of energy". Similar such modern, succinct definitions of heat are as follows:

- In a thermodynamic sense, heat is never regarded as being stored within a body. Like work, it exists only as *energy in transit* from one body to another; in thermodynamic terminology, between a system and its surroundings. When energy in the form of heat is added to a system, it is stored not as heat, but as kinetic and potential energy of the atoms and molecules making up the system
- The noun heat is defined only during the process of energy transfer by conduction or radiation
- Heat is defined as any spontaneous flow of energy from one object to another, caused by a difference in temperature between two objects
- Heat may be defined as *energy in transit* from a high-temperature object to a lower-temperature object
- Heat as an interaction between two closed systems without exchange of work is a pure heat interaction when the two systems, initially isolated and in a stable equilibrium, are placed in contact.
  The energy exchanged between the two systems is then called heat
- Heat is a form of energy possessed by a substance by virtue of the vibrational movement, i.e. kinetic energy, of its molecules or atoms
- Heat is the transfer of energy between substances of different temperatures

THERMODYNAMICS

INTERNAL ENERGY

Heat is related to the internal energy U of the system and work W done *by* the system by the first law of thermodynamics:

$$\Delta U = Q - W$$

which means that the energy of the system can change either via work or via heat flows across the boundary of the thermodynamic system. In more detail, Internal energy is the sum of all microscopic forms of energy of a system.

It is related to the molecular structure and the degree of molecular activity and may be viewed as the sum of kinetic and potential energies of the molecules; it comprises the following types of energies:

Type	*Composition of Internal Energy (U)*
Sensible energy	The portion of the internal energy of a system associated with kinetic energies (molecular translation, rotation, and vibration; electron translation and spin; and nuclear spin) of the molecules.
Latent energy	The internal energy associated with the phase of a system.
Chemical energy	The internal energy associated with the atomic bonds in a molecule
Nuclear energy	The tremendous amount of energy associated with the strong bonds within the nucleus of the atom itself.
Energy interactions	Those types of energies not stored in the system (e.g. heat transfer, mass transfer, and work), but which are recognized at the system boundary as they cross it, which represent gains or losses by a system during a process
Thermal energy	The sum of sensible and latent forms of internal energy.

The transfer of heat to an ideal gas at constant pressure increases the internal energy and performs boundary work (i.e. allows a control volume of gas to become larger or smaller), provided the volume is not constrained. Returning to the first law equation and separating the work term into two types, "boundary work" and "other" (e.g. shaft work performed by a compressor fan), yields the following:

$$\Delta U + W_{boundary} = Q - W_{other}$$

This combined quantity Ä$U + W_{boundary}$ is enthalpy, H, one of the thermodynamic potentials. Both enthalpy, H, and internal energy, U are state functions. State functions return to their initial values upon completion of each cycle in cyclic processes such as that of a heat engine.

In contrast, neither Q nor W are properties of a system and need not sum to zero over the steps of a cycle. The infinitesimal expression for heat, äQ, forms an inexact differential for processes involving work. However, for processes involving no change in volume, applied magnetic field, or other external parameters, äQ, forms an exact differential. Likewise, for adiabatic processes (no heat transfer), the expression for work forms an exact differential, but for processes involving transfer of heat it forms an inexact differential

HEAT CAPACITY

For a simple compressible system such as an ideal gas inside a piston, the changes in enthalpy and internal energy can be related to the heat capacity at constant pressure and volume, respectively. Constrained to have constant volume, the heat, Q, required to change its temperature from an initial temperature, T_0, to a final temperature, T_f is given by:

$$Q = \int_{T_0}^{T_f} C_v dT = \Delta U$$

Removing the volume constraint and allowing the system to expand or contract at constant pressure:

$$Q = \Delta U \int_{V_0}^{V_f} PdT = \Delta H = \int_{T_0}^{T_f} C_p dT$$

For incompressible substances, such as solids and liquids, the distinction between the two types of heat capacity disappears, as no work is performed. Heat capacity is an extensive quantity and as such is dependent on the number of molecules in the system. It can be represented as the product of mass, m , and specific heat capacity, c_s according to:

$$C_p = mc_s$$

or is dependent on the number of moles and the molar heat capacity, c_n according to:

$$C_p = nc_n$$

The molar and specific heat capacities are dependent upon the internal degrees of freedom of the system and not on any external properties such as volume and number of molecules.

The specific heats of monatomic gases (e.g., helium) are nearly constant with temperature. Diatomic gases such as hydrogen display some temperature dependence, and triatomic gases (e.g., carbon dioxide) still more in liquids at sufficiently low temperatures, quantum effects become significant. An example is the behaviour of bosons such as helium-4. For such substances, the behaviour of heat capacity with temperature is discontinuous at the Bose-Einstein condensation point

The quantum behaviour of solids is adequately characterized by the Debye model. At temperatures well below the characteristic Debye temperature of a solid lattice, its specific heat will be proportional to the cube of absolute temperature. For low-temperature metals, a second term is needed to account for the behaviour of the conduction electrons, an example of Fermi-Dirac statistics

CHANGES OF PHASE

The boiling point of water, at sea level and normal atmospheric pressure and temperature, will always be at nearly 100 °C, no matter how much heat is added.

The extra heat changes the phase of the water from liquid into water vapour.The heat added to change the phase of a substance in this way is said to be "hidden" and thus it is called latent heat (from the Latin *latere* meaning "to lie hidden"). Latent heat is the heat per unit mass necessary to change the state of a given substance, or:

$$L = \frac{Q}{\Delta m}$$

and

$$Q = \int_{M_0}^{M} L dm.$$

Note that, as pressure increases, the L rises slightly. Here, M_o is the amount of mass initially in the new phase, and M is the amount of mass that ends up in the new phase. Also,L generally does not depend on the amount of mass that changes phase, so the equation can normally be written:

$$Q = L\text{Ä}m.$$

Sometimes L can be time-dependent if pressure and volume are changing with time, so that the integral can be written as:

$$Q = \int L \frac{dm}{dt} dt.$$

HEAT TRANSFER MECHANISMS

Heat tends to move from a high-temperature region to a low-temperature region. This heat transfer may occur by the mechanisms of conduction and radiation. In engineering, the term *convective heat transfer* is used to describe the combined effects of conduction and fluid flow and is regarded as a third mechanism of heat transfer

Conduction

Conduction is the most significant means of heat transfer in a solid. On a microscopic scale, conduction occurs as hot, rapidly moving or vibrating atoms and molecules interact with neighboring atoms and molecules, transferring some of their energy (heat) to these neighboring atoms. In insulators the heat flux is carried almost entirely by phonon vibrations

The "electron fluid" of a conductive metallic solid conducts nearly all of the heat flux through the solid. Phonon flux is still present, but carries less than 1% of the energy. Electrons also conduct electric current through conductive solids, and the thermal and electrical conductivities of most metals have about the same ratio. A good electrical conductor, such as copper, usually also conducts heat well. The Peltier-Seebeck effect exhibits the propensity of electrons to conduct heat through an electrically conductive solid. Thermoelectricity is caused by the relationship between electrons, heat fluxes and electrical currents

Convection

Convection is usually the dominant form of heat transfer in liquids and gases. This is a term used to characterize the combined effects of conduction and fluid flow. In convection, enthalpy transfer occurs by the movement of hot or cold portions of the fluid together with heat transfer by conduction. Commonly an increase in temperature produces a reduction in density. Hence, when water is heated on a stove, hot water from the bottom of the pan rises, displacing the colder more dense liquid which falls. Mixing and conduction result eventually in a nearly homogenous density and even temperature. Two types of convection are commonly distinguished, *free convection*, in which gravity and buoyancy forces drive the fluid movement, and *forced convection*, where a fan, stirrer, or other means is used to move the fluid. Buoyant convection is due to the effects of gravity, and hence does not occur in microgravity environments

Radiation

Radiation is the only form of heat transfer that can occur in the absence of any form of medium; thus it is the only means of heat transfer through a vacuum. Thermal radiation is a direct result of the movements of atoms and molecules in a material. Since these atoms and molecules are composed of charged particles (protons and electrons), their movements result in the emission of electromagnetic radiation, which carries energy away from the surface. At the same time, the surface is constantly bombarded by radiation from the surroundings, resulting in the transfer of energy to the surface. Since the amount of emitted radiation increases with increasing temperature, a net transfer of energy from higher temperatures to lower temperatures results

The power that a black body emits at various frequencies is described by Planck's law. For any given temperature, there is a frequency f_{max} at which the power emitted is a maximum. Wien's displacement law, and the fact that the frequency of light is inversely proportional to its wavelength in vacuum, mean that the peak frequency f_{max} is proportional to the absolute temperature T of the black body. The photosphere of the Sun, at a temperature of approximately 6000 K, emits radiation principally in the visible portion of the spectrum. The earth's atmosphere is partly transparent to visible light, and the light reaching the earth's surface is absorbed or reflected. The earth's surface emits the absorbed radiation, approximating the behaviour of a black body at 300 K with spectral peak at f_{max}. At these lower frequencies, the atmosphere is largely opaque and radiation from the earth's surface is absorbed or scattered by the atmosphere. Though some radiation escapes into space, it is absorbed and subsequently re-emitted by atmospheric gases. It is this spectral selectivity of the atmosphere that is responsible for the planetary greenhouse effect.

The common household lightbulb has a spectrum overlapping the blackbody spectra of the sun and the earth. A portion of the photons emitted by a tungsten light bulb filament at 3000K are in the visible spectrum. However, most of the energy is associated with photons of longer wavelengths; these will not help a person, but will still transfer heat to the environment, as can be deduced empirically by observing a household incandescent lightbulb. Whenever EM radiation is emitted and then absorbed, heat is transferred. This principle is used in microwave ovens, laser cutting, and RF hair removal

Other Heat Transfer Mechanisms

- Latent heat: Transfer of heat through a physical change in the medium such as water-to-ice or water-to-steam involves significant energy and is exploited in many ways: steam engine, refrigerator etc.
- Heat pipes: Using latent heat and capillary action to move heat, heat pipes can carry many times as much heat as a similar-sized copper rod. Originally invented for use in satellites, they are starting to have applications in personal computers

HEAT DISSIPATION

In cold climates, houses with their heating systems form dissipative systems. In spite of efforts to insulate such houses to reduce heat losses to their exteriors, considerable heat is lost, or dissipated, from them, which can make their interiors uncomfortably cool or cold. For the comfort of its inhabitants, the interior of a house must be maintained out of thermal equilibrium with its external surroundings. In effect, domestic residences are oases of warmth in a sea of cold and the thermal gradient between the inside and outside is often quite steep.

This can lead to problems such as condensation and uncomfortable draughts (drafts) which, if left unaddressed, can cause structural damage to the property. This is why modern insulation techniques are required to reduce heat loss

In such a house, a thermostat is a device capable of starting the heating system when the house's interior falls below a set temperature, and of stopping that same system when another (higher) set temperature has been achieved. Thus the thermostat controls the flow of energy into the house, that energy eventually being dissipated to the exterior

TEMPERATURE MEASUREMENT

Temperature is the most commonly measured parameter, yet in many respects it is the least understood. It is a surprisingly difficult parameter to measure with the precision that one might reasonably expect.

To obtain accuracies better than 0.2°C (0.4°F) great care in needed. Errors occur due to the presence of temperature gradients, drafts, sensor nonlinearities, poor thermal contact, calibration drifts, radiant energy and sensor self heating. Generally the accuracy of all sensor types can be greatly improved by individual calibration.

The information in this section is oriented towards electronic thermometers those with an electrical output that can be connected to a measuring instrument, such as: a data acquisition system, a data logger, a control system or a chart recorder. However, there is also a wide range of thermometers that can be used for manual temperature measurement. These include: the glass thermometer, various gas thermometers, pressure based thermometers, bimetallic thermometers and temperature sensitive paint or film thermometers.

IS TEMPERATURE MEASUREMENT DIFFICULT?

The answer depends on the temperature, the material being measured and your expectations of accuracy. The table below summarises the difficulty of temperature measurement over a range of temperatures:

Accuracy Required

Temperature	*±5°C*	*±1°C*	*±0.5°C*	*±0.1°C*
-200°C	care needed	difficult	difficult	very difficult
0°C to 50°C	easy	care needed	difficult	very difficult
1000°C	care needed	very difficult	extremely difficult	almost impossible
2000°C	very difficult	extremely difficult	almost impossible	don't even try

In a laboratory with appropriate standards and equipment, it is possible to measure temperature to 0.001°C (1°mC) or even better. This is typically done by interpolation (estimation of the values) between two standards, using a quality platinum temperature sensor and / or a Type S thermocouple.

When measuring temperature it is important to keep your goals in mind. Identify exactly what is to be measured and the accuracy needed. If accurate temperature differences are of prime importance, then consider using the thermopile to avoid the need for closely matched sensors

SOURCES OF TEMPERATURE MEASUREMENT ERROR

In using temperature sensors it is helpful to think of where heat flows. This applies to both sheathed and unsheathed sensors. Understanding the thermal resistances and where they are located is especially useful in identifying potential errors sources.

The following diagram indicates some of the complexity in temperature measurement. Note the presence of thermal gradients in the material being measured. These gradients can be particularly troublesome when measuring the materials with poor thermal conductivity, such as plastics and even stainless steel

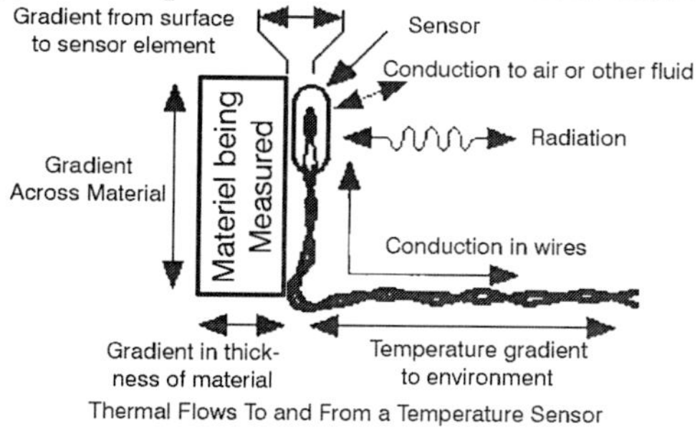

Thermal Flows To and From a Temperature Sensor

Below is a description of temperature measurement error sources and some suggestions on minimising these errors.

SENSOR CALIBRATION

Sensors calibration errors can be due to offset, scale and linearity errors. In addition, each of these errors can drift with time and temperature cycling. Hysteresis (where a value depends on the direction from which it was approached) can be noticed with some sensors, but the effect is usually small with the exception of the bimetallic strip where it may be several degrees. Platinum RTD's are considered the most accurate and stable of standard sensors. However, individually calibrated thermocouples can come close over the same temperature range. The platinum based thermocouples can be just a stable as platinum RTD's and cover a higher temperature range

Sensor interchangeability is often the decisive factor. It refers to the maximum temperature reading error likely to occur in replacing a sensor with another of the same type without recalibrating the system. Choosing a practical calibration reference can be an issue. For professional purposes, a high quality platinum RTD is best, along with an appropriate indicator. Other references include iced water bath, traditional glass thermometers (especially laboratory grade) and medical thermometers. In general, the defining reference points of the ITS-90 are not practical for routine calibration purposes

THERMAL GRADIENTS

Thermal gradients are often a major source of measurement error. This is especially true when measuring materials with poor thermal conductivity such as: air, most liquids and non-metallic solids. In the case of fluids it important that the fluid be stirred. An unstirred ice bath (a mixture of ice and water) can have a vertical temperature gradient of several degrees. If stirring is not practical, gradients can be minimised by insulating the system being measured, to prevent heat transfer into or out of, the system. Employing multiple sensors for spatial diversity and averaging the readings is another solution

HEAT CONDUCTION IN SENSOR LEADS

All sensors with the exception of non-contact and maybe the fibre optic types require that wires be brought to the sensor. These wires are usually copper, an excellent heat conductor. The placement of these wires can have a significant impact on accuracy.

The wires allow heat flow into or out of the sensor body, requiring the sensing element to be in better thermal contact with the material being measured than would otherwise be needed. When measuring the temperature of thermal insulation materials, this can be a major source of error.

There are three solutions, all of which are good standard practice:

- Use as thin wires as is practical for sensor hook-up. (Note: this contradicts good practice for high temperature thermocouple measurement where the reverse rule applies-use the thickest wire that is practical)
- Place the wires in or against the material being measured, so that they actually assist in transferring the temperature of the material into the sensor

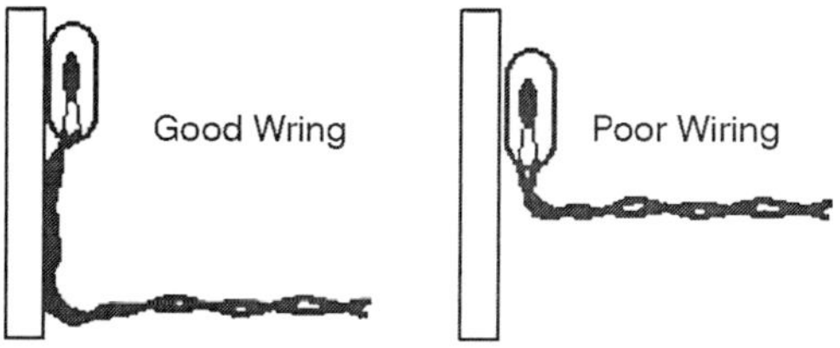

Heat Conductance in leads

- Attempt to minimise the thermal gradient along the sensor wires by placing the wires at an angle to the gradient. This ensures a higher thermal resistance because of a longer length of wire

RADIATION

Radian heat can be a major source of error in measuring air temperature.

A sensor in sunlight is almost certain to read significantly higher than the actual air temperature.

To avoid this error the sensor must be shielded from source of radiant energy.

The sun is the most obvious source, however just about any object that is at a different temperature to the air is a potential source (or sink) of troublesome radiant energy

The best solutions are the following:

- Ensure that the sensor's outside surface is highly reflective of infrared radiation (i.e.it is painted white or has a bright metal finish)

- Ensure the sensor is thermally 'well connected' to the air by having a good surface area-to-volume ratio. Small sensors are generally better.
- Place the sensor in a vented radiation shield that also has a highly reflective surface on the outside and inside
- Ensure the sensor has a high surface area-to-volume ratio to ensure good thermal 'contact' with the air

Radiant heat loss can be a source of sensor error when measuring elevated temperature. Again, the same rules apply. Use reflective surface finish on the sensor, shield the sensor if possible, and ensure a good thermal contact with the medium being measured

SENSOR SELF-HEATING

Thermistors, RTDs and semiconductor sensors require the application of an excitation power in order that a reading may be taken. This power can heat the sensor, causing a high reading. The effect depends on the size of the sensing element and the level of power. Typically, the magnitude of the self heating effect is between 0.1°C and 1.5°C.

The best solutions are the following:

- Calibrate out the self-heating effect. This is perhaps the easiest solution. However, the equipment must be allowed time to 'warm up', and different calibrations are required for mediums with different thermal characteristics e.g. air and water
- Use the lowest possible excitation power. However, a compromise between self-heating and sensitivity (and signal-to-noise ratio) must be made
- Avoid unnecessarily small sensing elements-they will self-heat more than larger elements
- Switch the excitation power off between readings. This is the best solution if the readings can be made quickly, before the sensor has time to warm, and if there is adequate time between readings for cooling
- Avoid self-heating sensor types-use thermocouples. However this is not necessary as simple as that, as the measuring device is likely to use a reference junction temperature sensor that is itself prone to self-heating.

THERMAL CONTACT

Obviously, thermal contact with the material being measured is important, but the degree of contact required is dependent on other parasitic thermal connections to the surroundings which are likely to have a significant impact on heat flow. These parasitics include: lead conduction, direct contact with other material (e.g. air) and radiant energy transfers

If there is no temperature gradient in the vicinity of the sensor, the thermal contact of the sensor can be poor and the sensor will still provide accurate readings

THERMAL TIME CONSTANT

When the temperature changes, it takes time for a sensor to respond. Some sensors respond quickly, some in less than a second, while others take minutes or even hours.

The time taken to reach 63% of the way to the new temperature is referred to as the 'thermal time constant'. Most sensors have one dominant time constant. However, sometimes, there are minor, but longer, time constants present that can confuse the measurement process.

Obviously if the temperature is changing more quickly than a sensor is able to track, the measurement will be in error.

The best solutions include the following:

- Use a more rapidly responding sensor
- Improve thermal contact
- Reduce the sensors thermal mass, by minimising material in contact with the sensing element that is not associated with improving thermal contact
- Compensate the readings using an inverse matching filter. If the thermal characteristics of the system are constant and known, it is possible to predict the temperature dynamically.

Sometimes long time constants are useful in providing an averaging effect on a rapidly fluctuating temperature. If this effect is to be exploited, care needs to be taken to compensate for the phase (time) delay in the response

READ-OUT ERRORS

The measuring device connected to the sensor is never perfect. The measuring device, be it a: meter, chart recorder, data acquisition board or data logger, can have calibration, linearity and temperature dependent errors

These errors can be reduced by:

- Calibration of the readout device against known references
- Calibration of the total sensor - readout system, using a reference temperature or against a precision thermometer (glass or RTD+readout)

Temperature effects on the readout device can be a subtle source of error. It is recommended that a test be conducted where the temperature of the sensor is held constant, but the readout device be placed in an oven or freezer. This is particularly important for thermocouple read-out devices, as their performance can be greatly impacted by temperature gradients and the quality of the internal reference junction sensing

ELECTRICAL NOISE OR INTERFERENCE

Electrical noise can induce errors in systems with poor noise rejection. Use the standard procedures as described in the Measurements Methods section to minimise the impact. These include the following:

- Use shielded twisted pair cable
- Keep wiring away from power cables, transformers and electrical machinery
- Install low pass filters into the measuring device
- Avoid ground loops

In some industrial processes, electrical noise can be so intense that non-contact or fibre-optic sensors are the only option.

CONDENSATION

Sometimes in situations where temperatures are frequently cycled through the dew point, condensation in the sensor and wiring can collect and become an electrical leakage path, causing errors. Prevention is better than cure.

Preventative measures include the following:

- Ensure that the sensor and its wiring is sealed. When an air volume is sealed, ensure that the air is dry
- Moisture can wick along wires by capillary action. This can be inhibited by sealing wires at the sensor end, with a medium-viscosity super glue, or using a penetrating oil from both ends.

Condensation can be a source of intermittent errors and may go unrecognized for a long time. It also can lead to corrosion, accelerated by sensor excitation power, ultimately leading to complete failure. Semiconductor sensors can be particularly prone to moisture penetrating the metal-plastic interface of plastic packages. Evaporating condensation can also lead to measurement errors due to evaporative cooling effects - a subtle but real error source.

SENSOR MECHANICAL STRESS

Many temperature sensor elements can respond to mechanical stress. film type RTD has the appearance of a strain gauge and will behave like one given the opportunity. Again, prevention is better than cure

*Preventative measures in*clude:

- Ensure that sensing elements are not subjected to deformation in the way they are mounted
- Avoid using adhesives in attaching sensors to the surfaces to be measured. The differences in the thermal coefficients of linear expansion will induce mechanical stress
- Use sensor that are less sensitive to stress - the thermocouple
- Wound (as opposed to film) RTDs can be prone to vibration damage. Take great care in the selection and mounting of sensors in high-vibration environment
- Use grease in preference to adhesive to ensure thermal connection. Avoid potting sensor elements in epoxy

Calibration

If funds are available, special purpose temperature calibrators can be purchased. However, these are expensive and not always as accurate as one might expect

To calibrate over the normal environment range, there are two low cost standards that can be effectively employed to achieve 0.2°C accuracy. These are ice-water mix made from distilled water and a standard medical thermometer. These points (0.0°C and about 37°C) are sufficiently far apart to provide a useful calibration over the -20°C to +60°C range

REFERENCES

- A Guide to Physics: Thermodynamics, Statistical Physics, and Quantum Mechanics by Gerald D. Mahan, Boris E. Nadgorny, and Max Dresden.
- The Language of Science by Sidney B. Cahn.
- Introduction to Thermal Sciences: Thermodynamics Fluid Dynamics Heat Transfer by Frank W. Schmidt, Robert E. Henderson, and Carl H. Wolgemuth.
- Thermodynamics In Materials Science by Robert T. Dehoff.
- Heat and Thermodynamics: A Historical Perspective (Greenwood Guides to Great Ideas in Science) by Christopher J. T. Lewis.

2

Heat Transfer

CONDUCTIVE HEAT TRANSFER

HEAT TRANSFER MODES

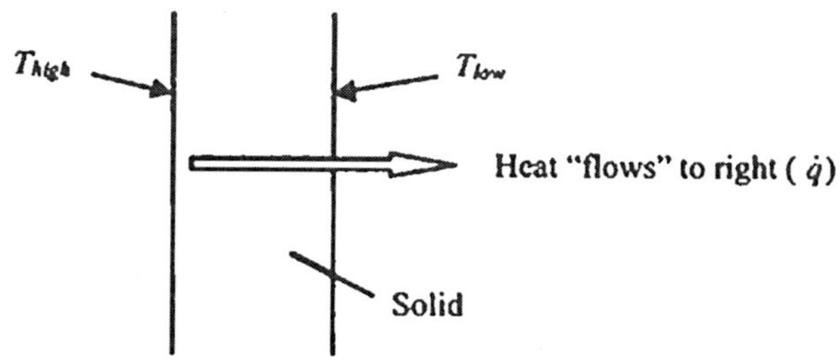

Fig. Conduction Heat Transfer

Heat transfer processes are classified into three types. The first is conduction, which is defined as transfer of heat occurring through intervening matter without bulk motion of the matter. The process pictorially. A solid (a block of metal, say) has one surface at a high temperature and one at a lower temperature. This type of heat conduction can occur, through a turbine blade in a jet engine. The outside surface, which is exposed to gases from the combustor, is at a higher temperature than the inside surface, which has cooling air next to it. The level of the wall temperature is critical for a turbine blade.

The second heat transfer process is convection, or heat transfer due to a flowing fluid. The fluid can be a gas or a liquid; both have applications in aerospace technology. In convection heat transfer, the heat is moved through bulk transfer of a non-uniform temperature fluid.

The third process is radiation or transmission of energy through space without the necessary presence of matter. Radiation is the only method for heat transfer in space. Radiation can be important even in situations in which there is an intervening medium; a familiar example is the heat transfer from a glowing piece of metal or from a fire.

Conduction

We will start by examining conduction heat transfer. We must first determine how to relate the heat transfer to other properties (either mechanical, thermal, or geometrical). The answer to this is rooted in experiment, but it can be motivated by considering heat flow along a "bar" between two heat reservoirs at T_A, T_B. It is plausible that the heat transfer rate, Q, is a function of the temperature of the two reservoirs, the bar geometry and the bar properties. (Are there other factors that should be considered? If so, what?). This can be expressed as

$$\dot{Q} = f_1(T_A, T_B, \text{bar geometry, bar properties}).$$

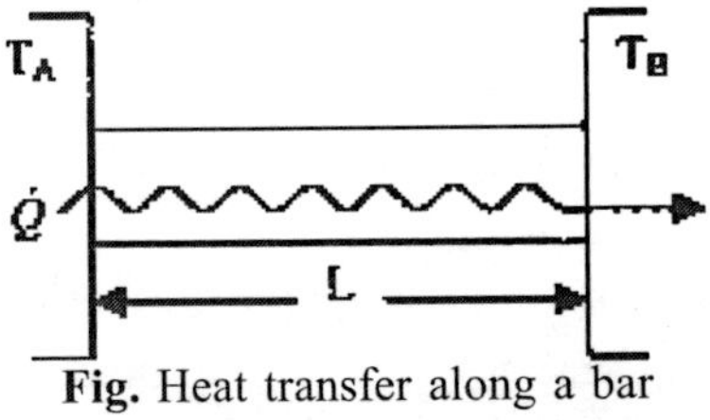

Fig. Heat transfer along a bar

It also seems reasonable to postulate that $\dot{Q}$ should depend on the temperature difference $T_A - T_B$. $T_A - T_B$ If is zero, then the heat transfer should also be zero. The temperature dependence can therefore be expressed as $\dot{Q} = f_2\ [T_A - T_B, T_A$, bar geometry, bar properties].

An argument for the general form of f_2 can be made from physical considerations. One requirement, as said, is if $f_2 = 0$. $T_A = T_B$ Using a Mac Laurin series expansion, as follows,

$$f(\Delta T) = f(0) + \left.\frac{\partial f}{\partial(\Delta T)}\right|_0 \Delta T + ...$$

If we define $\Delta T = T_A - T_B$ and $f = f_2$, we find that (for small $T_A - T_B$), $f_2(T_A - T_B) = \dot{Q} = f_2(0) + \left.\frac{\partial f_2}{\partial(T_A - T_B)}\right|_{T_A - T_B = 0} (T_A - T_B) + ...$

We know that $f_2(0) = 0$. The derivative evaluated at $T_A = T_B$ (thermal equilibrium) is a measurable property of the bar. In addition, we know that $\dot{Q} > 0$ if TA>TB or $\partial f_2/\partial(T_A - T_B) > 0$. It also seems reasonable that if we had two bars of the same area, we would have twice the heat transfer, so that we can postulate that $\dot{Q}$ is proportional to the area.

Finally, although the argument is by no means rigorous, experience leads us to believe that as L increases $\dot{Q}$ should get smaller. All of these lead to the generalization that, for the bar, the derivative in Equation has the form

$$\left.\frac{\partial f_2}{\partial(T_A - T_B)}\right|_{T_A - T_B = 0} = \frac{kA}{L}.$$

In Equation, k is a proportionality factor that is a function of the material and the temperature, $A$is the cross-sectional area and $L$is the length of the bar. In the limit for any temperature difference ΔT across a length Δx as both L, $T_A - T_B \rightarrow 0$, we can say

$$\dot{Q} = kA\frac{(T_A - T_B)}{L} = -kA\frac{(T_B - T_A)}{L} = -kA\frac{dT}{dx}.$$

A more useful quantity to work with is the heat transfer per unit area, defined as

$$\dot{q} \equiv \frac{\dot{Q}}{A}.$$

The quantity $\dot{q}$ is called the heat flux and its units are Watts/ m^2. The expression in can be written in terms of heat flux as

$$\dot{q} = -k\frac{dT}{dx}.$$

Equation is the one-dimensional form of Fourier's law of heat conduction. The proportionality constant k is called the thermal conductivity. Its units are W/m-K. Thermal conductivity is a well-tabulated property for a large number of materials. Some values for familiar materials are given in Table; others can be found in the references. The thermal conductivity is a function of temperature and the values shown in Table are for room temperature.

Table: Thermal conductivity at room temperature for some metals and non-metals (W/m-K)

Metals	*Ag*	*Cu*	*Al*	*Fe*	*Steel*		
	420	390	200	70	50		
Non-metals	H_2O	Air	Engine oil	H2	Brick	Wood	Cork
	0.6	0.026	0.15	0.18	0.4-0.5	0.2	0.04

Steady-State One-Dimensional Conduction

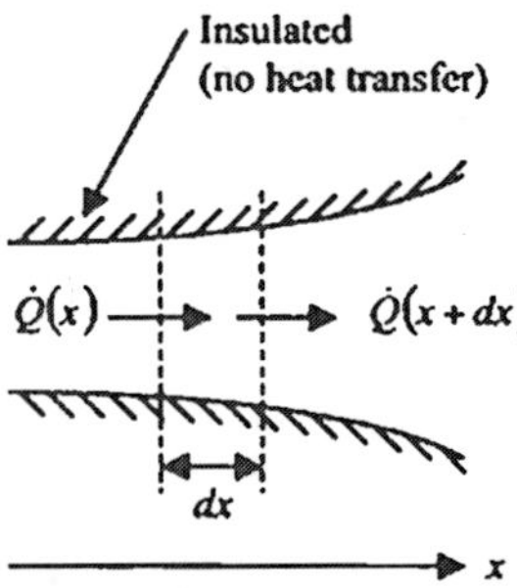

Fig. One-dimensional Heat Conduction

For one-dimensional heat conduction (temperature depending on one variable only), we can devise a basic description of the process. The first law in control volume form (steady flow energy equation) with no shaft work and no mass flow reduces to the statement that $\Sigma\dot{Q}$ for all surfaces = 0 (no heat transfer on top or bottom). From Equation, the heat transfer rate in at the left (at x) is

$$\dot{Q}(x) = -k\left(A\frac{dT}{dx}\right)_x.$$

The heat transfer rate on the right is

$$\dot{Q}(x+dx) = \dot{Q}(x) + \left.\frac{d\dot{Q}}{dx}\right|_x dx + ...$$

Using the conditions on the overall heat flow and the expressions

$$\dot{Q}(x) - \left(\dot{Q}(x) + \frac{d\dot{Q}}{dx}(x)dx + ...\right) = 0.$$

Taking the limit as dx approaches zero we obtain

$$\frac{d\dot{Q}(x)}{dx} = 0,$$

or

$$\frac{d}{dx}\left(kA\frac{dT}{dx}\right) = 0.$$

If is constant (i.e. if the properties of the bar are independent of temperature), this reduces to

$$\frac{d}{dx}\left(A\frac{dT}{dx}\right) = 0,$$

or (using the chain rule)

$$\frac{d^2T}{dx^2} + \left(\frac{1}{A}\frac{dA}{dx}\right)\frac{dT}{dx} = 0.$$

Equation or describes the temperature field for quasi-one-dimensional steady state (no time dependence) heat transfer. We now apply this to an example.

Heat Transfer Through a Plane Slab

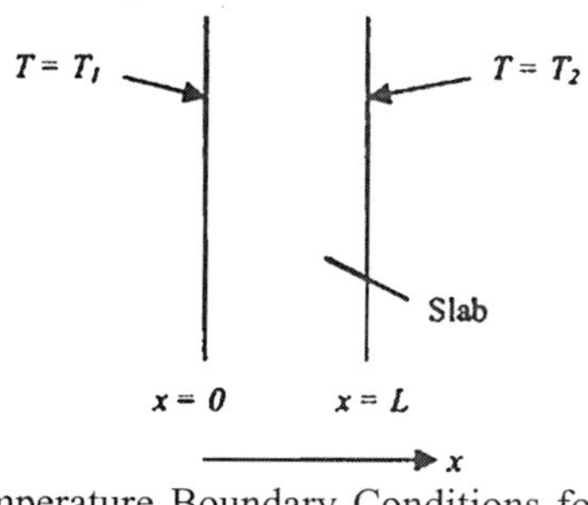

Fig. Temperature Boundary Conditions for a Slab

For this configuration, the area is not a function of x, i.e. A. Equation, thus becomes

$$\frac{d^2T}{dx^2} = 0.$$

Equation, can be integrated immediately to yield

$$\frac{dT}{dx} = a$$

and

$$T = ax + b.$$

Equation is an expression for the temperature field where a and b are constants of integration. For a second order equation, such as, we need two boundary conditions to determine and . One such set of boundary conditions can be the specification of the temperatures at both sides of the slab $T(0) = T_1$; $T(L) = T_2$.

The condition $T(0) = T_1$ implies that $b = T_1$. The condition $T_2 = T(L)$ implies that $T_2 = aL + T_1$, or $a = (T_2 - T_1)/L$. With these expressions for and the temperature distribution can be written as

$$T(x) = T_1 + \left(\frac{T_2 - T_1}{L}\right)x.$$

This linear variation in temperature for a situation in which $T_1 > T_2$.

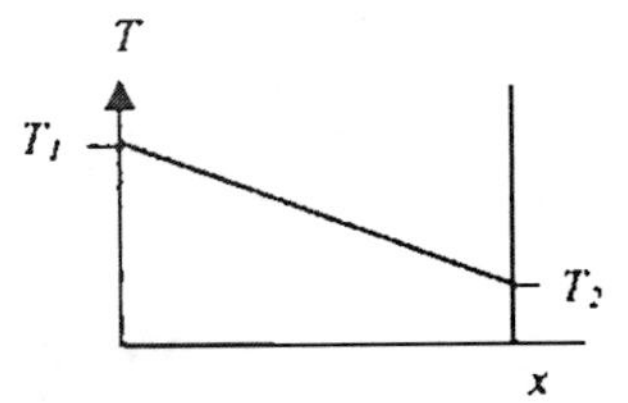

Fig. Temperature Distribution Through a Slab

The heat flux $\dot{q}$ is also of interest. This is given by

$$\dot{q} = -k\frac{dT}{dx} = -k\frac{(T_2 - T_1)}{L} = \text{constant}.$$

Thermal Resistance Circuits

There is an electrical analogy with conduction heat transfer that can be exploited in problem solving. The analog of $\dot{Q}$ is current, and the analog of the temperature difference, $T_1 - T_2$, is voltage difference. From this perspective the slab is a pure resistance to heat transfer and we can define

$$\dot{Q} = \frac{T_1 - T_2}{R}$$

where $R = L/kA$, the thermal resistance. The thermal resistance Rincreases as Lincreases, as Adecreases, and as $k$decreases.

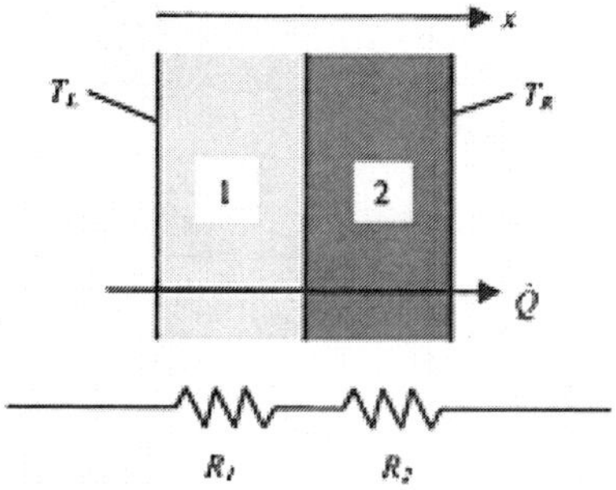

Fig. Heat Transfer Across a Composite Slab (Series Thermal Resistance)

The concept of a thermal resistance circuit allows ready analysis of problems such as a composite slab (composite planar heat transfer surface). In the composite slab, the heat flux is constant with . The resistances are in series and sum to $R = R_1$ R_2. If T_L is the temperature at the left, and T_R is the temperature at the right, the heat transfer rate is given by

$$\dot{Q} = \frac{T_L - T_R}{R} = \frac{T_L - T_R}{R_1 + R_2}.$$

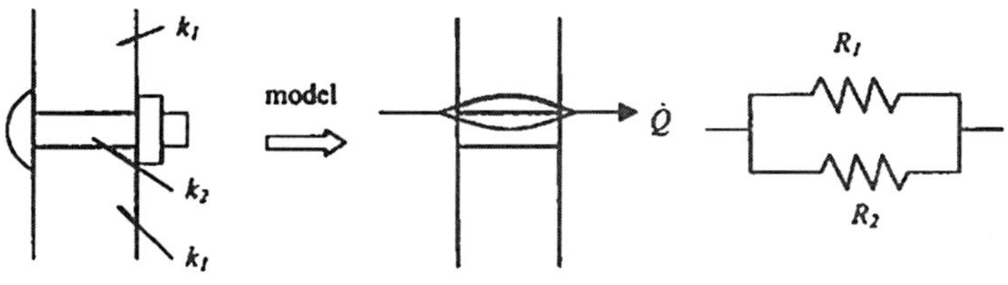

Fig. Heat Transfer for a Wall with Dissimilar Materials (Parallel Thermal Resistance)

Another example is a wall with a dissimilar material such as a bolt in an insulating layer. In this case, the heat transfer resistances are in parallel. The physical configuration, the heat transfer paths and the thermal resistance circuit.

For this situation, the total heat flow $\dot{Q}$ is made up of the heat flow in the two parallel paths, $\dot{Q} = \dot{Q}_1 + \dot{Q}_2$, with the total resistance given by

$$\frac{1}{R} = \frac{1}{R_1} + \frac{1}{R_2}.$$

More complex configurations can also be examined; a brick wall with insulation on both sides.

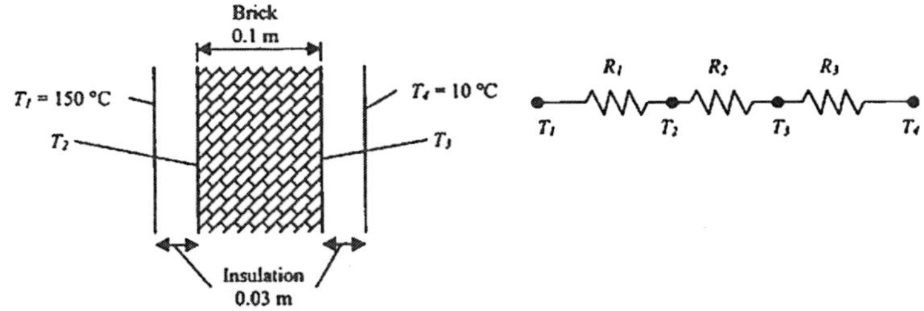

Fig. Heat Transfer Through an Insulated Wall

The overall thermal resistance is given by

$$R = R_1 + R_2 + R_3 = \frac{L_1}{k_1 A_1} + \frac{L_2}{k_2 A_2} + \frac{L_3}{k_3 A_3}.$$

Some representative values for the brick and insulation thermal conductivity are:

$$k_{\text{brick}} = k_2 = 0.7 \text{ W/m-K}$$
$$k_{\text{insulation}} = k_1 = k_3 = 0.07 \text{ W/m-K}$$

Using these values, and noting that $A_1 = A_2 = A_3 = A$, we obtain

$$AR_1 = AR_3 = \frac{L_1}{k_1} = \frac{0.03 \text{ m}}{0.07 \text{ W/m-K}} = 0.42 \text{ m}^2 \text{ K/W}$$

$$AR_2 = \frac{L_2}{k_2} = \frac{0.1 \text{ m}}{0.7 \text{ W/m-K}} = 0.14 \text{ m}^2 \text{ K/W}.$$

This is a series circuit so

$$\dot{q} = \frac{\dot{Q}}{A} = \text{constant throughout} = \frac{T_1 - T_4}{RA} = \frac{140 \text{ K}}{0.98 \text{ m}^2 \text{ K/W}} = 142 \text{ W/m}^2.$$

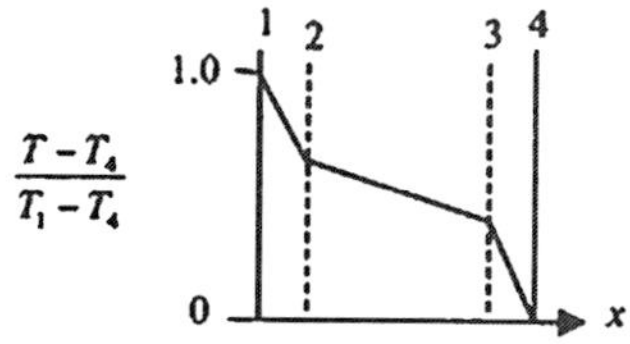

Fig. Temperature Distribution Through an Insulated Wall

The temperature is continuous in the wall and the intermediate temperatures can be found from applying the resistance equation across each slab, since $\dot{Q}$ is constant across the slab. to find T_2:

$$\dot{q} = \frac{T_1 - T_2}{R_1 A} = 142 \text{ W/m}^2.$$

This yields $T_1 - T_2 = 60$ K or $T_2 = 90$°C.

The same procedure gives $T_3 = 70$°C. As sketched, the larger drop is across the insulating layer even though the brick layer is much thicker.

Steady Quasi-One-Dimensional Heat Flow in Non-Planar Geometry

The quasi one-dimensional equation that has been developed can also be applied to non-planar geometries, such as cylindrical and spherical shells.

Cylindrical Shell

An important case is a cylindrical shell, a geometry often encountered in situations where fluids are pumped and heat is transferred.

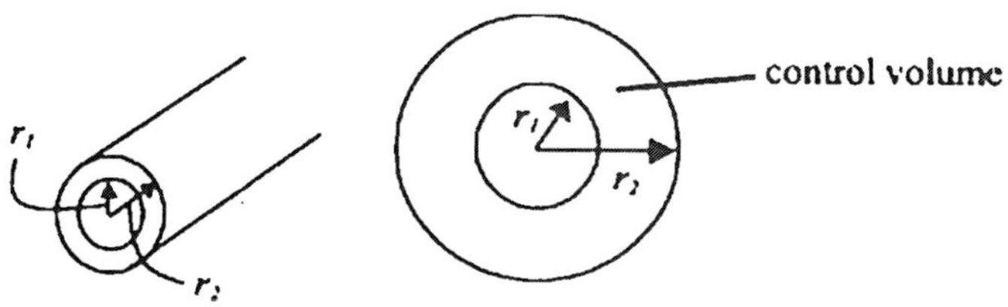

Fig. Cylindrical Shell Geometry Notation

For a steady axisymmetric configuration, the temperature depends only on a single coordinate (r) and Equation, can be written as

$$k = \frac{d}{dr}\left(A(r)\frac{dT}{dr}\right) = 0,$$

or, since $A = 2\pi r$,

$$\frac{d}{dr}\left(r\frac{dT}{dr}\right) = 0.$$

The steady-flow energy equation (no fluid flow, no work) tells us that $\dot{Q}_{\text{in}} = \dot{Q}_{\text{out}}$, or

$$\frac{d\dot{Q}}{dr} = 0.$$

The heat transfer rate per unit length is given by

$$\dot{Q} = -k2\pi r\frac{dT}{dr}.$$

Equation is a second order differential equation for T. Integrating this equation once gives

$$r\frac{dT}{dr} = a,$$

where a is a constant of integration. Equation, can be written as

$$dT = a\frac{dr}{r}$$

where both sides of Equation are exact differentials. It is useful to cast this equation in terms of a dimensionless normalized spatial variable so we can deal with quantities of order unity. To do this, divide through by the inner radius, r_1,

$$dT = a\frac{d(r/r_1)}{(r/r_1)}.$$

Integrating, yields

$$T = a\ln\frac{(r)}{(r_1)} + b.$$

To find the constants of integration a and b, boundary conditions are needed. These will be taken to be known temperatures T_1 and T_2 at r_1 and r_2 respectively. Applying $T = T_1$ $r = r_1$ at gives $T_1 = b$. Applying $T = T_2$ at $r = r_2$ yields

$$T_2 = a \ln\left(\frac{r_2}{r_1}\right) + T_1,$$

or

$$a = \frac{T_2 - T_1}{\ln(r_2 / r_1)}.$$

The temperature distribution is thus

$$T = (T_2 - T_1)\frac{\ln(r / r_1)}{\ln(r_2 / r_1)} + T_1.$$

As said, it is generally useful to put expressions into non-dimensional and normalized form so that we can deal with numbers of order unity (this also helps in checking whether results are consistent). If convenient, having an answer that goes to zero at one limit is also useful from the perspective of ensuring the answer makes sense. Equation can be put in nondimensional form as

$$\frac{T - T_1}{T_2 - T_1} = \frac{\ln(r / r_1)}{\ln(r_2 / r_1)}.$$

The heat transfer rate, $\dot{Q}$, is given by

$$\dot{Q} = -kA\frac{dT}{dr} = -2\pi r_1 k\frac{(T_2 - T_1)}{\ln(r_2 / r_1) r_1}\frac{1}{r_1} = 2\pi\frac{k(T_1 - T_2)}{\ln(r_2 / r_1)}$$

per unit length. When the heat flow rate is written so as to incorporate our definition of thermal resistance,

$$\dot{Q} = \frac{T_1 - T_2}{R},$$

comparison with, reveals the thermal resistance R to be

$$R = \frac{\ln (r_1 / r_1)}{2\pi k}.$$

The cylindrical geometry can be viewed as a limiting case of the planar slab problem. To make the connection, consider the case when $(r_2 - r_1)/r_1 \ll 1$. From the series expansion for $\ln (1 + x)$ we recall that

$$\ln (1 + x) \approx x - \frac{x^2}{2} + \frac{x^3}{3} + \ldots$$

(Look it up, try it numerically, or use the binomial theorem on the series ($\frac{1}{1+x} = 1 - x + x^2 + ...$) and integrate term by term.)

The logarithms in Equation, can thus be written as

$$\ln\left(1 + \frac{r - r_1}{r_1}\right) \cong \frac{r - r_1}{r_1}$$

and

$$\ln\left(\frac{r_2}{r_1}\right) \cong \frac{r_2 - r_1}{r_1}$$

in the limit of $(r_2 - r_1) \ll r_1$. Using these expressions in Equation, gives

$$T = T_2 - T_1 \frac{(r - r_1)}{(r_2 - r_1)} + T_1.$$

With the substitution of $r - r_1 = x$, and $r_2 - r_1 = L$ we obtain

$$T = T_1 + (T_2 - T_1)\frac{x}{L},$$

which is the same as Equation. The plane slab is thus the limiting case of the cylinder if $(r_2 - r_1)/R_1 \ll 1$, where the heat transfer can be regarded as taking place in (approximately) a planar slab.

To see when this is appropriate, consider the expansion ln $(1 + x)/x$, which is the ratio of heat flux for a cylinder and a plane slab.

For <10% error, the ratio of thickness to inner radius should be less than 0.2, and for 20% error, the thickness to inner radius should be less than 0.5.

Table: Utility of plane slab approximation

x	0.1	0.2	0.3	0.4	0.5
$\frac{\ln(1+x)}{x}$	0.95	0.91	0.87	0.84	0.81

Spherical Shell

A second example is the spherical shell with specified temperatures $T(r_1) = T_1$ and $T(r_2) = T_2$.

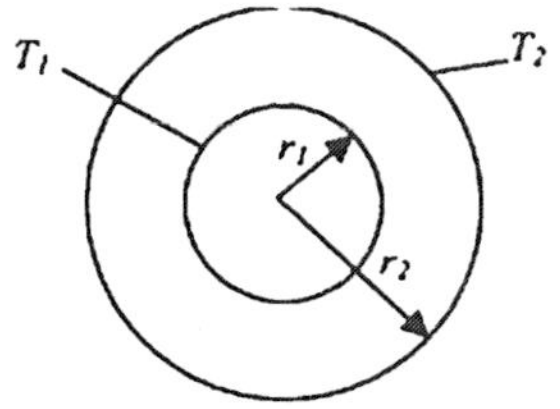

Fig. Spherical Shell

The area is now $A(r) = 4\pi r^2$, so the equation for the temperature field is

$$\frac{d}{dr}\left(r^2 \frac{dT}{dr}\right) = 0.$$

Integrating Equation, once yields

$$\frac{dT}{dr} = \frac{a}{r^2}.$$

Integrating again gives

$$T = -\frac{a}{r} + b,$$

or, normalizing the spatial variable

$$T = \frac{a'}{(r/r_1)} + b,$$

where a' and b are constants of integration. As before, we specify the temperatures at $r = r_1$ and $r = r_2$. Use of the first boundary condition gives $T(r_1) = T_1 = a' + b$. Applying the second boundary condition gives

$$T(r_2) = T_2 \frac{a'}{(r_2/r_1)} + b.$$

Solving for a' and b,

$$a' = \frac{T_1 - T_2}{1 - r_1/r_2}$$

$$b' = T_1 - \frac{T_1 - T_2}{1 - r_1/r_2}.$$

In non-dimensional form the temperature distribution is thus

$$\frac{T_1 - T}{T_1 - T_2} = \frac{1 - (r_1/r)}{1 - (r_1/r_2)}.$$

CONVECTIVE HEAT TRANSFER

The secocnd type of heat transfer to be examined is convection, where a key problem is determining the boundary conditions at a surface exposed to a flowing fluid. An example is the wall temperature in a turbine blade because turbine temperatures are critical for creep (and thus blade) life. A view of the problem is given which shows a cross-sectional view of a turbine blade.

There are three different types of cooling indicated, all meant to ensure that the metal is kept at a temperature much lower than that of the combustor exit flow in which the turbine blade operates. In this case, the turbine wall temperature is not known and must be found as part of the solution to the problem.

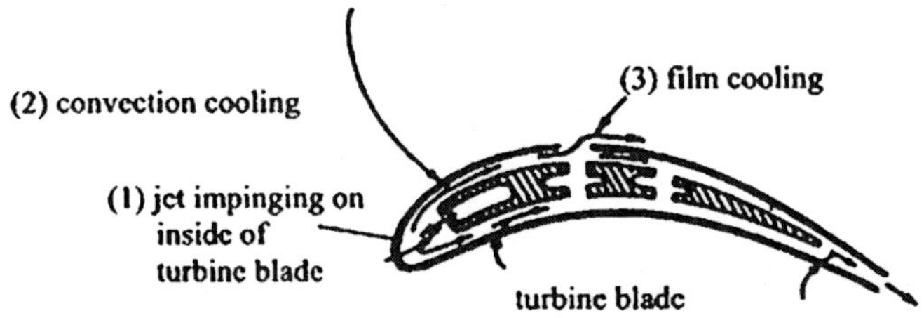

Fig. Turbine Blade Heat Transfer Configuration

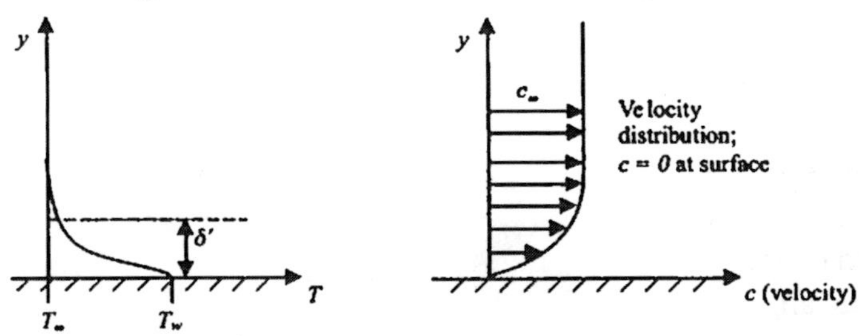

Fig. Temperature and Velocity Distributions near a Surface

To find the turbine wall temperature, we need to analyze convective heat transfer, which means we need to examine some features of the fluid motion near a surface. The conditions near a surface are illustrated schematically. In a region of thickness δ, there is a thin "film" of slowly moving fluid through which most of the temperature difference occurs. Outside this layer, T is roughly uniform (this defines δ′). The heat flux can thus be expressed as

$$\dot{q} = \frac{\dot{Q}}{A} = \frac{k(T_w - T_\infty)}{\delta'}.$$

It cannot be emphasized enough that this is a very crude picture. The general concept, however, is correct, in that close to the wall, there is a thin layer in which heat is transferred basically by conduction. Outside of this region is high mixing. The difficulty is that the thickness of the layer is not a fluid property.

It depends on velocity (Reynolds number), structure of the wall surface, pressure gradient and Mach number. Generally δ' is not known and needs to be found and it is customary to calculate the heat transfer using k_{fluid}/δ'.

This quantity has the symbol h and is known as the convective heat transfer coefficient. The units of h are W/m^2K. The convective heat transfer coefficient is defined by

$$\dot{q} = \frac{\dot{Q}}{A} = h(T_w - T_\infty).$$

Equation is often called Newton's Law of Cooling. For many situations of practical interest, the quantity is still known mainly through experiments.

THE REYNOLDS ANALOGY

We describe the physical mechanism for the heat transfer coefficient in a turbulent boundary layer because most aerospace vehicle applications have turbulent boundary layers. The treatment closely follows that in Eckert and Drake. Very near the wall, the fluid motion is smooth and laminar, and molecular conduction and shear are important. The shear stress, τ, at a plane is given by $\mu = \frac{dc}{dy} = \tau$ (where μ is the dynamic viscosity), and the heat flux by $\dot{q} = -k\frac{dT}{dy}$. The latter is the same expression that was used for a solid. The boundary layer is a region in which the velocity is lower than the free stream. In a turbulent boundary layer, the dominant mechanisms of shear stress and heat transfer change in nature as one moves away from the wall.

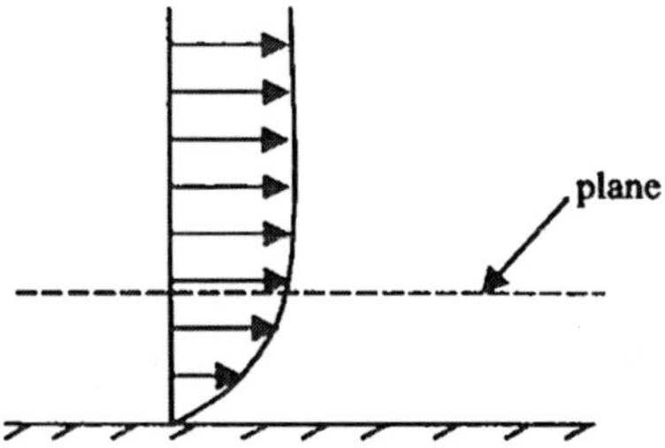

Fig. Velocity Profile Near a Surface

As one moves away from the wall (but still in the boundary layer), the flow is turbulent. The fluid particles move in random directions and the transfer of momentum and energy is mainly through interchange of fluid particles.

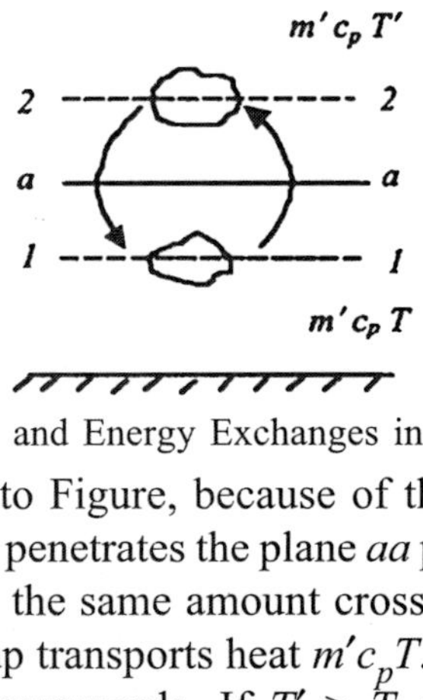

Fig. Momentum and Energy Exchanges in Turbulent Flow

With reference to Figure, because of the turbulent velocity field, a fluid mass m' penetrates the plane *aa* per unit time and unit area. In steady flow, the same amount crosses *aa* from the other side. Fluid moving up transports heat $m'c_pT$. Fluid moving down transports $m'c_pT'$ downwards. If $T' > T$, there is a turbulent downwards heat flow $\dot{q}_{turbulent}$, given by $\dot{q}_{turbulent} = m'c_p(T' - T)$, that results.

Fluid moving up also has momentum $m'c$ and fluid moving down has momentum $m'c'$. The net flux of momentum down per unit area and time is therefore $m'(c' - c)$. This net flux of momentum per unit area and time is a force per unit area or stress, given by

$$\tau_{turbulent} = m'(c' - c).$$

Based on these considerations, the relation between heat flux and shear stress at plane *aa* is

$$\dot{q}_{\text{turbulent}} = \tau_{\text{turbulent}}\, c_p \left(\frac{T' - T}{c' - c}\right),$$

or (again approximately)

$$\dot{q}_{\text{turbulent}} = \tau_{\text{turbulent}}\, c_p \left(\frac{dT}{dc}\right),$$

since the locations of planes 1-1 and 2-2 are arbitrary.

For the laminar region, the heat flux towards the wall is $\dot{q} = -kdT/dy$ and dividing by the expression for the shear stress, $\tau = \mu dc/dy$, yields

$$\dot{q} = -\tau \frac{k}{\mu}\frac{dT}{dc}.$$

The same relationship is applicable in laminar or turbulent flow if $k/\mu = c_p$ or, expressed slightly differently,

$$\frac{\mu c_p}{k} \frac{\mu/\rho}{k/\rho c_p} = \frac{v}{\alpha} = 1,$$

where v is the kinematic viscosity, and α is the thermal diffusivity.

The quantity $\mu c_p/k$ is known as the Prandtl number (Pr), after the man who first presented the idea of the boundary layer and was one of the pioneers of modern fluid mechanics. For gases, Prandtl numbers are in fact close to unity and for air $Pr = 0.71$ at room temperature. The Prandtl number varies little over a wide range of temperatures: approximately 3% from 300-2000 K.

We want a relation between the values at the wall (at which $T = T_w$ and $c = 0$) and those in the free stream. To get this, we integrate the expression for dT from the wall to the free stream

$$-dT = \frac{1}{c_p}\frac{\dot{q}}{\tau} dc,$$

where the relation between heat transfer and shear stress has been taken as the same for both the laminar and the turbulent portions of the boundary layer. The assumption being made is that the mechanisms of heat and momentum transfer are similar. Equation can be integrated from the wall to the freestream (conditions "at ∞"):

$$-\int_{w}^{\infty} dT = \frac{1}{c_p}\int_{w}^{\infty}\left(\frac{\dot{q}}{\tau}\right)dc,$$

where $\dot{q}/\tau$ and c_p are assumed constant.

Carrying out the integration yields

$$T_w - T_\infty = \frac{\dot{q}_w c_\infty}{\tau_w c_p},$$

where $\mathbf{c}_\infty$ is the velocity and c_p is the specific heat. In Equation, $\dot{q}_w$ is the heat flux to the wall and τ_w is the shear stress at the wall. The relation between skin friction (shear stress) at the wall and heat transfer is thus

$$\frac{\dot{q}_w}{\rho_\infty c_p (T_w - T_\infty)\, c_\infty} = \frac{\tau_w}{\rho_\infty c_\infty^2}.$$

The quantity

$$\frac{\tau_w}{\frac{1}{2}\rho_\infty c_\infty^2}$$

is known as the skin friction coefficient and is denoted by C_f. The skin friction coefficient has been tabulated (or computed) for a large number of situations. If we define a non-dimensional quantity

$$\frac{\dot{q}_w}{\rho_\infty c_p (T_w - T_\infty) c_\infty} = \frac{h(T_\infty - T_w)}{\rho_\infty c_p (T_w - T_\infty) c_\infty} = \frac{h}{\rho_\infty c_p c_\infty} = St,$$

known as the Stanton Number, we can write an expression for the heat transfer coefficient, *h* as

$$h \approx \rho_\infty c_p c_\infty \frac{C_f}{2}.$$

Equation provides a useful estimate of *h*, or $\dot{q}_w$, based on knowing the skin friction, or drag. The direct relationship between the Stanton Number and the skin friction coefficient is

$$St = \frac{C_f}{2}.$$

The relation between the heat transfer and the skin friction coefficient

$$\dot{q}_w \approx \frac{\tau_w c_p (T_w - T_\infty)}{c_\infty}$$

is known as the Reynolds analogy between shear stress and heat transfer. The Reynolds analogy is extremely useful in obtaining a first approximation for heat transfer in situations in which the shear stress is "known."

An example of the use of the Reynolds analogy is in analysis of a heat exchanger. One type of heat exchanger has an array of tubes with one fluid flowing inside and another fluid flowing outside, with the objective of transferring heat between them. To begin, we need to examine the flow resistance of a tube. For fully developed flow in a tube, it is more appropriate to use an average velocity $\overline{c}$ and a bulk temperature T_B. Thus, an approximate relation for the heat transfer is

$$\dot{q}_w \approx \tau_w c_p \frac{T_B - T_w}{\overline{c}}.$$

The fluid resistance (drag) is all due to shear forces and is given by $\tau_w A_w = D$, where A_w is the tube "wetted" area (perimeter length). The total heat transfer, $\dot{Q}$, is $\dot{q}_w A_w$, so that

$$\dot{Q} = D c_p \frac{T_B - T_w}{\overline{c}}.$$

The power, P, to drive the flow through a resistance is given by the product of the drag and the velocity, $D\overline{c}$, so that

$$\frac{\dot{Q}}{P} = \frac{c_p (T_B - T_w)}{\overline{c}^2}.$$

The mass flow rate is given by $\dot{m} = \rho \overline{c} A$, where is the cross sectional area. For a given mass flow rate and overall heat transfer rate, the power scales as $\overline{c}^2$ or as $1/A^2$, i.e.,

$$P \propto \frac{\dot{Q}\dot{m}^2}{\rho^2 c_p (T_B - T_w)} \frac{1}{A^2}.$$

Equations shows that to decrease the power dissipated, wc need to decrease $\overline{c}$, which can be accomplished by increasing the cross-sectional area. Two possible heat exchanger configurations the one on the right will have a lower loss.

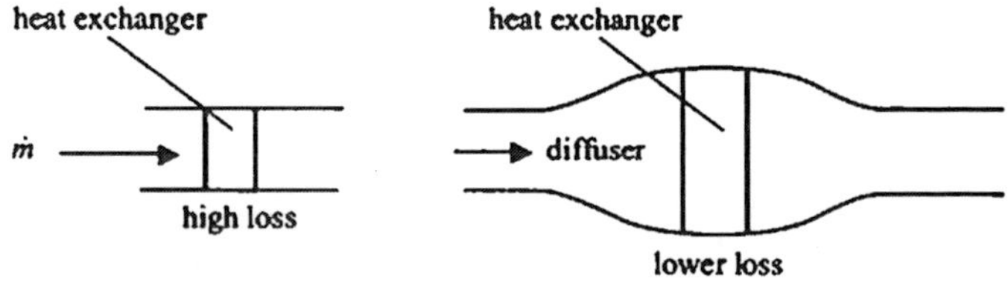

Fig. Heat Exchanger Configurations

To recap, there is an approximate relation between skin friction (momentum flux to the wall) and heat transfer called the Reynolds analogy that provides a useful way to estimate heat transfer rates in situations in which the skin friction is known. The relation is expressed by

$$St = \frac{C_f}{2},$$

or

$$\frac{\text{heat flux to wall}}{\text{convected heat flux}} = \frac{\text{momentum flux to wall}}{\text{convected momentum flux}},$$

or

$$\frac{\dot{q}_w}{\rho_\infty c_\infty c_p (T_\infty - T_w)} = \frac{\tau_w}{\rho_\infty c_\infty^2}.$$

The Reynolds analogy can be used to give information about scaling of various effects as well as initial estimates for heat transfer. It is emphasized that it is a useful tool based on a hypothesis about the mechanism of heat transfer and shear stress and not a physical law.

COMBINED CONDUCTION AND CONVECTION

We can now analyze problems in which both conduction and convection occur, starting with a wall cooled by flowing fluid on each side. A description of the convective heat transfer can be given explicitly as

$$\frac{\dot{Q}}{A} = \dot{q} = h\,(T_w - T_\infty).$$

This could represent a model of a turbine blade with internal cooling.

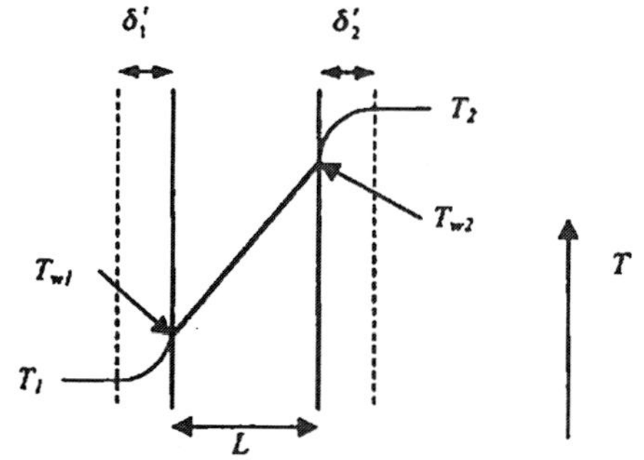

Fig. Conducting Wall with Convective Heat Transfer

The heat transfer in fluid 1 is given by

$$\frac{\dot{Q}}{A} = h_1 \; (T_{w1} - T_1).$$

which is the heat transfer per unit area to the fluid. The heat transfer in fluid 2 is similarly given by

$$\frac{\dot{Q}}{A} = h_2 \; (T_2 - T_{w2}).$$

Across the wall, we have

$$\frac{\dot{Q}}{A} = \frac{k}{L} (T_{w2} - T_{w1}).$$

The quantity $\dot{Q}/A$ is the same in all of these expressions. Putting them all together to write the known overall temperature drop yields a relation between heat transfer and overall temperature drop, $T_2 - T_1$:

$$T_2 - T_1 = (T_2 - T_{w/2}) + (T_{w2} - Tw1) + (T_{w1} - T_1) = \frac{\dot{Q}}{A}\left[\frac{1}{h_1} + \frac{L}{k} + \frac{1}{h_2}\right].$$

We can define a thermal resistance, R, as before, such that

$$\dot{Q} = \frac{(T_2 - T_1)}{R},$$

where R is given by

$$R = \frac{1}{h_1 A} + \frac{L}{AK} + \frac{1}{h_2 A}.$$

Equation is the thermal resistance for a solid wall with convection heat transfer on each side.

For a turbine blade in a gas turbine engine, cooling is a critical consideration. In terms of Figure, T_2 is the combustor exit (turbine inlet) temperature and T_1 is the temperature at the compressor exit. We wish to find T_{w2} because this is the highest metal temperature. From, the wall temperature can be written as

$$T_{w2} = T_2 - \frac{\dot{Q}}{Ah_2} = T_2 - \frac{T_2 - T_1}{R}\frac{1}{Ah_2}.$$

Using the expression for the thermal resistance, the wall temperatures can be expressed in terms of heat transfer coefficients and wall properties as

$$T_{w2} = T_2 - \frac{T_2 - T_1}{\frac{h_2}{h_1} + \frac{Lh_2}{k} + 1}.$$

Equation provides some basic design guidelines. The goal is to have a low value of T_{w2}. This means h_1/h_2 should be large, k should be large (but we may not have much flexibility in choice of material) and L should be small. One way to achieve the first of these is to have h_2 low.

A second example of combined conduction and convection is given by a cylinder exposed to a flowing fluid.

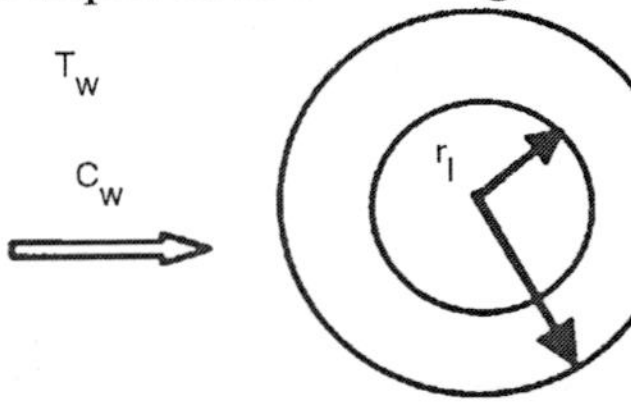

Fig. Cylinder in a Flowing Fluid

For the cylinder the heat flux at the outer surface is given by

$$\dot{q} = \frac{\dot{Q}}{A} = h(T_w - T_\infty) \text{ at } r = r_2.$$

The boundary condition at the inner surface could be either a heat flux condition or a temperature specification; we use the latter to simplify the algebra. Thus, $T = T_1$ at $r = r_1$. This is a model for the heat transfer in a pipe of radius r_1 surrounded by insulation of thickness $r_2 - r_1$. The solution for a cylindrical region was given as

$$T(r) = a \ln\left(\frac{r}{r_1}\right) + b.$$

Use of the boundary condition $T(r_1) = T_1$ yields $b = T_1$.

At the interface between the cylinder and the fluid, $r = r_2$, the temperature and the heat flow are continuous.

$$\dot{q} = - \underbrace{-k\frac{dT}{dr}}_{\substack{\text{heart flux}\\ \text{inside cylinder}}} = -k\frac{a}{r_2} = \underbrace{h\left[\left(a\ln\left(\frac{r_2}{r_1}\right) + T_1\right) - T_\infty\right]}_{\text{surface heat flux to flu id}}$$

Plugging the form of the temperature distribution in the cylinder into Equation yields

$$-a\left(\frac{k}{r_2} + h\ln\left(\frac{r_2}{r_1}\right)\right) = h(T_1 - T_\infty).$$

The constant of integration, a, is

$$a = \frac{-h(T_1 - T_\infty)}{\frac{k}{r_2} + h\ln\left(\frac{r_2}{r_1}\right)} = -\frac{(T_1 - T_\infty)}{\frac{k}{hr_2} + \ln\left(\frac{r_2}{r_1}\right)}$$

and the expression for the temperature is, in normalized non-dimensional form,

$$\frac{T_1 - T}{T_1 - T_\infty} = \frac{\ln(r/r_1)}{\frac{k}{hr_2} + \ln(r_2/r_1)}.$$

The heat flow per unit length, $\dot{Q}$, is given by

$$\dot{Q} = \frac{2\pi(T_1 - T_\infty)k}{\frac{k}{hr_2} + \ln(r_2/r_1)}.$$

The units in Equation are W/m-s.

A problem of interest is choosing the thickness of insulation to minimize the heat loss for a fixed temperature difference $T_1 - T_\infty$ between the inside of the pipe and the flowing fluid far away from the pipe. ($T_1 - T_\infty$ is the driving temperature distribution for the pipe.) To understand the behaviour of the heat transfer we examine the denominator in Equation as r_2 varies. The thickness of insulation that gives maximum heat transfer is given by

$$\frac{d}{dr_2}\left(\frac{k}{hr_2}+\ln\left(\frac{r_2}{r_1}\right)\right)=0.$$

From Equation, the value of r_2 for maximum $\dot{Q}$ is thus

$$(r_2)_{\text{maximum heat transfer}}=\frac{k}{h}.$$

If r_2 is less than this, we can add insulation and increase heat loss. To understand why this occurs, which shows a schematic of the thermal resistance and the heat transfer. As r_2 increases from a value less than $r_2 = k/h$, two effects take place. First, the thickness of the insulation increases, tending to drop the heat transfer because the temperature gradient decreases.

Secondly, the area of the outside surface of the insulation increases, tending to increase the heat transfer. The second of these is (loosely) associated with the k/hrs_2 term, the first with the ln (r_2/r_1) term. There are thus two competing effects which combine to give a maximum $\dot{Q}$ at $r_2 = k/h$.

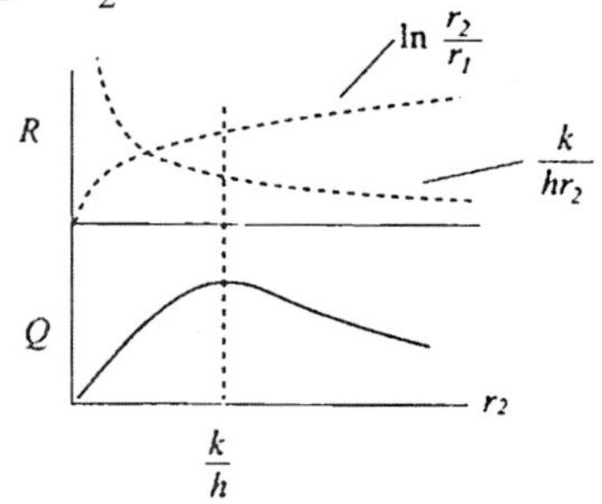

Fig. Critical Radius of Insulation

DIMENSIONLESS NUMBERS AND ANALYSIS OF RESULTS

Phenomena in fluid flow and heat transfer depend on dimensionless parameters. The Mach number and the Reynolds number are two you have already seen. These parameters give information as to the relevant flow regimes of a given solution. Casting equations in dimensionless form helps show the generality of application to a broad class of situations (rather than just one set of dimensional parameters).

It is generally good practice to use non-dimensional numbers, forms of equations, and results presentation whenever possible. The results for heat transfer from the cylinder are already in dimensionless form but we can carry the idea even further. For the cylinder, we had in Equation,

$$\frac{T-T_1}{T_\infty-T_1}=\frac{\ln(r/r_1)}{\dfrac{k}{hr_2}+\ln(r_2/r_1)}.$$

The parameter hr_2/k or hL/k, where L is a relevant length for the particular problem of interest, is called the Biot number, denoted by Bi. In terms of this parameter,

$$\frac{T-T_1}{T_\infty-T_1}=\frac{\ln(r/r_1)}{\dfrac{1}{Bi}+\ln(r_2/r_1)}.$$

The size of the Biot number gives a key to the regimes in which different features are dominant. For Bi $\gg$ 1 the convection heat transfer process offers little resistance to heat transfer. There is thus only a small ΔT outside (i.e. $T(r_2)$ close to T_∞) compared to the ΔT through the solid with a limiting behaviour of

$$\frac{T-T_1}{T_\infty-T}=\frac{\ln(r/r_1)}{\ln(r_2/r_1)}$$

as Bi goes to infinity. This is much like the situation with an external temperature specified.

For Bi $\ll$ 1the conduction heat transfer process offers little resistance to heat transfer. The temperature difference in the body (i.e. from $r_1$to r_2) is small compared to the external temperature difference, $T_1 - T_\infty$. In this situation, the limiting case is

$$\frac{T-T_1}{T_\infty-T_1}=\text{Bi}\ln\left(\frac{r}{r_1}\right).$$

In this regime there is approximately uniform temperature in the cylinder. The size of the Biot number thus indicates the regimes where the different effects become important.

RADIATION HEAT TRANSFER

All bodies radiate energy in the form of photons moving in a random direction, with random phase and frequency. When radiated photons reach another surface, they may either be absorbed, reflected or transmitted. The behaviour of a surface with radiation incident upon it can be described by the following quantities:

- α = absorptance - fraction of incident radiation absorbed
- ρ = reflectance - fraction of incident radiation reflected
- τ = transmittance - fraction of incident radiation transmitted.

Figure shows these processes graphically.

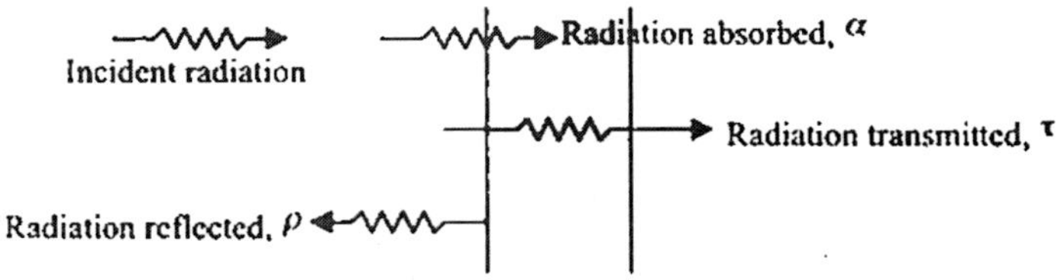

Fig. Radiation Surface Properties

From energy considerations the three coefficients must sum to unity

$$\alpha + \rho + \tau = 1.$$

Reflective energy may be either diffuse or specular (mirror-like). Diffuse reflections are independent of the incident radiation angle. For specular reflections, the reflection angle equals the angle of incidence.

IDEAL RADIATORS

An ideal thermal radiator is called a "black body." It has several properties:

- It has $\alpha = 1$, and absorbs all radiation incident on it.
- The energy radiated per unit area is $E_b = \sigma T^4$ where σ is the Stefan-Boltzmann constant,

$$\sigma = 5.67 \times 10^{-8} \text{ W/m}^2\text{K}^4.$$

The units of E_b are therefore W/m^2

The energy of a black body, E_b, is distributed over a range of wavelengths of radiation. We can define $e_\lambda = dE_b/d\lambda \approx \Delta E_b/\Delta\lambda$, the energy radiated per unit area for a range of wavelengths of width $\Delta\lambda$. The behaviour of $e\lambda$

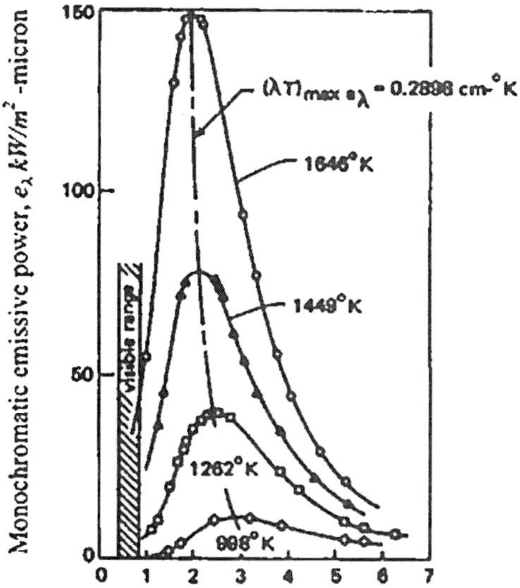

Fig. Emissive Power of a Black Body at Several Temperatures, Predicted and Observed; $(\lambda T)e\lambda_{max} = 0.2898$ cm K

The distribution of $e\lambda$ varies with temperature. The quantity λT at the condition where $e\lambda$ is a maximum is given by $(\lambda T)e\lambda_{max} = 0.2898$ cm K. As T increases, the wavelength for maximum energy emission shifts to shorter values. The frequency of the radiation, f, is given by $f = c/\lambda$ so high energy means short wavelengths and high frequency.

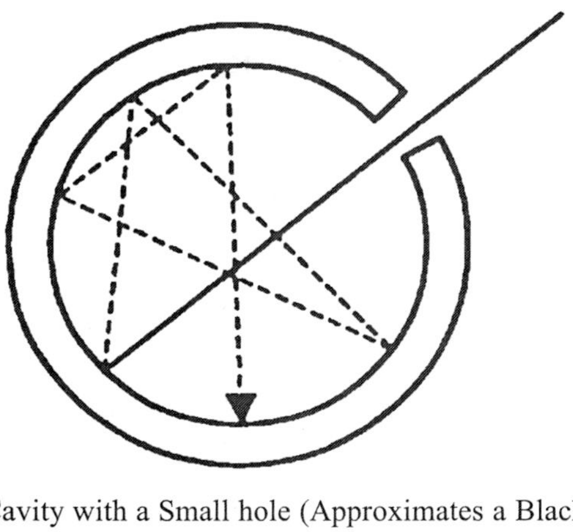

Fig. A Cavity with a Small hole (Approximates a Black Body)

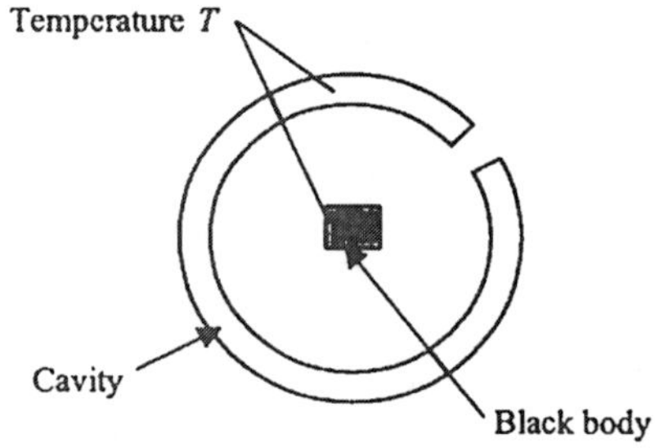

Fig. A Small Black Body Inside a Cavity

A physical realization of a black body is a cavity with a small hole. There are many reflections and absorptions. Very few entering photons (light rays) will get out. The inside of the cavity has radiation which is homogeneous and isotropic (the same in any direction, uniform everywhere).

Suppose we put a small black body inside the cavity. The cavity and the black body are both at the same temperature.

The radiant energy absorbed by the black body per second and per m^2 is $\alpha_B H$, where H is the irradiance, the radiant energy falling on any surface inside the cavity. The radiant energy emitted by the black body is E_B. Since $\alpha B = 1$ for a black body, $H = E_B$. The irradiance within a cavity whose walls are at temperature T is therefore equal to the radiant emittance of a black body at the same temperature and irradiance is a function of temperature only.

KIRCHHOFF'S LAW AND "REAL BODIES"

Real bodies radiate less effectively than black bodies. The measurement of this is the emittance, ε, defined by

$$\text{Emittance}: \varepsilon = \frac{E}{E_b},$$

where $E$is radiation from the real body at T, and E_b is radiation from a black body at T.

Values of emittance vary greatly for different materials. They are near unity for rough surfaces such as ceramics or oxidized metals, and roughly 0.02 for polished metals or silvered reflectors.

The level of the emittance can be related to the absorptance using the following arguments. Suppose we have a small non-black body in the cavity. The power absorbed per unit area is equal to

αH. The power emitted is equal to E. An energy balance gives $E = E_b\varepsilon = \alpha H = \alpha E_b$. Thus

$$\frac{E}{E_b} = \alpha = \varepsilon.$$

Equation, the relation $\alpha = \varepsilon$, is known as Kirchhoff's Law. It implies that good radiators are good absorbers. It was derived for the case when $T_{\text{body}} = T_{\text{surroundings (cavity)}}$ and is not strictly true for all circumstances when the temperature of the body and the cavity are different, but it is true if $\alpha_\lambda = \alpha$, $\varepsilon_\lambda = \varepsilon$, so the absorptance and emittance are not functions of λ. This situation describes a "gray body." Also, since α_λ, ε_λ are properties of the surface, $\alpha_\lambda = \varepsilon_\lambda$.

Radiation Heat Transfer Between Planar Surfaces

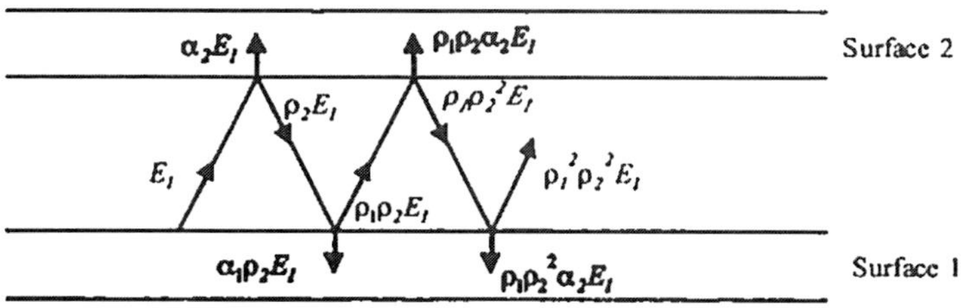

Fig. Path of a Photon between Two Gray Surfaces

Consider the two infinite gray surfaces. We suppose that the surfaces are thick enough so that $\alpha + \rho = 1$ (no radiation transmitted so = 0). Consider a photon emitted from Surface 1 (remembering that the reflectance $\rho = 1 - \alpha$):

Surface 1 emits E_1
Surface 2 absorbs $E_1\,\alpha_2$
Surface 2 reflects $E_1\,(1 - \alpha_2)$
Surface 1 absorbs $E_1\,(1 - \alpha_2)\,\alpha_1$
Surface 1 reflects $E_1\,(1 - \alpha_2)(1 - \alpha_1)$
Surface 2 absorbs $E_1\,(1 - \alpha_2)\,(1 - \alpha_1)\,\alpha_2$
Surface 2 reflects $E_1\,(1 - \alpha_2)\,(1 - \alpha_1)\,(1 - \alpha_2)$
Surface 1 absorbs $E_1(1 - \alpha_2)(1 - \alpha_1)(1 - \alpha_2)\,\alpha_1$

The same can be said for a photon emitted from Surface 2:

Surface 2 emits E_2
Surface 1 absorbs $E_2\alpha_1$

Surface 1 reflects $E_2(1 - \alpha_1)$
Surface 2 absorbs $E_2(1 - \alpha_2)\alpha_2$
Surface 2 reflects $E_2(1 - \alpha_1)(1 - \alpha_2)$

We can add up all the energy E_1 absorbed in 1 and all the energy E_2 absorbed in 2. In doing the bookkeeping, it is helpful to define $\beta = (1 - \alpha_1)(1 - \alpha_2)$. The energy E_1 absorbed in 1 is

$$E_1(1-\alpha_2)\alpha_1 + E_1(1-\alpha_2)\alpha_1(1-\alpha_2)(1-\alpha_1)+....$$

This is equal to

$$E_1(1-\alpha_2)\alpha_1(1+\beta+\beta^2+....).$$

However

$$\frac{1}{1-\beta} = (1-\beta)^{-1} = 1+\beta+\beta^2+....$$

We thus observe that the radiation absorbed by surface 1 can be written as

$$\frac{E_1(1-\alpha_2)\alpha_1}{1-\beta}.$$

Likewise

$$\frac{E_1(1-\alpha_1)\,\alpha_2}{1-\beta}$$

is the radiation generated at 2 and absorbed there as well. Putting this all together we find that

$$E_2 - \left(\frac{E_2(1-\alpha_1)\,\alpha_2}{1-\beta}\right) = \frac{E_2\alpha_1}{1-\beta}$$

is absorbed by 1. The net heat flux from 1 to 2 is

$$\begin{aligned}\dot{q}_{\text{net 1 to 2}} &= E_1 - \frac{E_1(1-\alpha_2)\,\alpha_1}{1-\beta} = \frac{E_2\alpha_1}{1-\beta} \\ &= \frac{E_1 - E_1(1-\alpha_1-\alpha_2+\alpha_1\alpha_2) - E_1\alpha_1 + E_1\alpha_1\alpha_2 - E_2\alpha_1}{1-(1-\alpha_1-\alpha_2+\alpha_1\alpha_2)}\end{aligned}$$

or

$$\dot{q}_{\text{net 1 to 2}} = \frac{E_1\alpha_2 - E_2\alpha_1}{\alpha_1+\alpha_2-\alpha_1\alpha_2}.$$

If $T_1 = T_2$, we would have $\dot{q} = 0$, so from Equation,

$$\frac{E_1}{\alpha_1} = \frac{E_2}{\alpha_2} = f(T).$$

If body 2 is black, $\alpha_2 = 1$, and $E_2 = \sigma T^4$.

$$\frac{E_1}{\alpha_1} = \sigma T^4,$$

$$\frac{\varepsilon_1 \sigma T^4}{\alpha_1} = \sigma T^4.$$

Therefore, again, $\varepsilon_1 = \alpha_1$ for any gray surface (Kirchhoff's Law).

Using Kirchhoff's Law we find,

$$\dot{q}_{\text{net 1 to 2}} = \frac{\varepsilon_1 \sigma T_1^4 \varepsilon_2 - \varepsilon_2 \sigma T_2^4 \varepsilon_1}{\varepsilon_1 + \varepsilon_2 - \varepsilon_1 \varepsilon_2},$$

or, as the final expression for heat transfer between gray, planar, surfaces,

$$\dot{q}_{\text{net 1 to 2}} = \frac{s(T_1^4 - T_2^4)}{\frac{1}{\varepsilon_1} + \frac{1}{\varepsilon_2} - 1}.$$

USE OF A THERMOS BOTTLE TO REDUCE HEAT TRANSFER

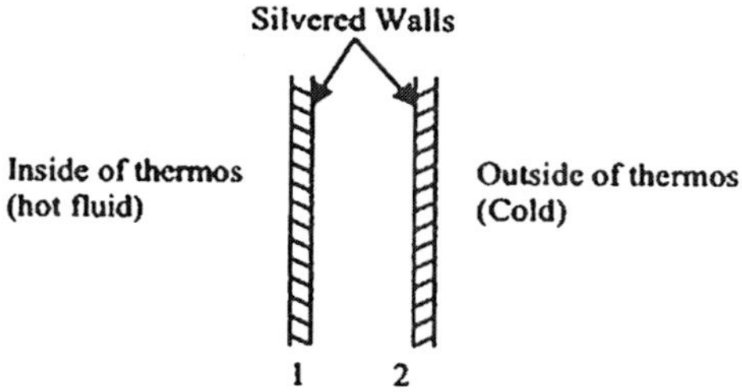

Fig. Schematic of a Thermos Wall

$\varepsilon_1 = \varepsilon_2 = 0.02$ for silvered walls. $T_1 = 100°\text{C} = 373$ K $T_2 = 20°\text{C} =$ 293K.

$$\dot{q}_{\text{net 1 to 2}} = \frac{\sigma(T_1^4 - T_2^4)}{\frac{1}{\varepsilon_1} + \frac{1}{\varepsilon_2} - 1} = \dot{q}_{\text{net 1 to 2}}$$

$$= \frac{(5.67 \times 10^{-8}\ \text{W/m}^2\text{K}^4)((373\ \text{K})^4 - (293\ \text{K})^4)}{\frac{1}{0.02} + \frac{1}{0.02} - 1} = 6.9\ \text{W/m}^2.$$

For the same ΔT, if we had cork insulation with $k = 0.04$ W/m-K, what thickness would be needed?

$$\dot{q} = \frac{k\Delta T}{L} \text{ so a thickness } L = \frac{k\Delta T}{L} = \frac{(0.04 \text{ W/m-K})(80 \text{ K})}{6.9 \text{ W/m}} = 0.47 \text{ m}$$

would be needed! The thermos is indeed a good insulator.

TEMPERATURE MEASUREMENT ERROR DUE TO RADIATION HEAT TRANSFER

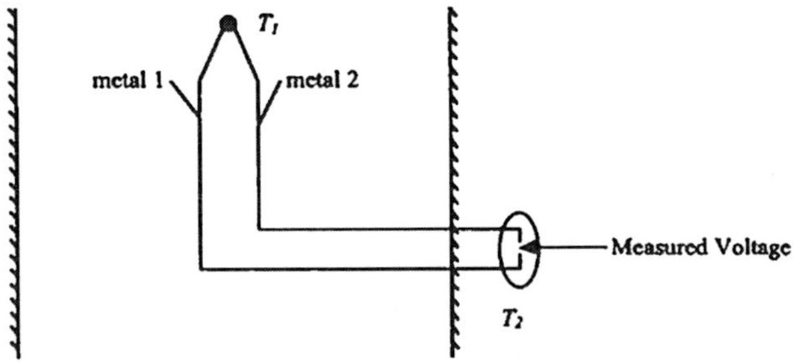

Fig. Thermocouple Used to Measure Temperature. Note: The Measured Voltage is Related to the Difference between T_1 and T_2 (the Latter is a known Temperature).

Thermocouples are commonly used to measure temperature. There can be errors due to heat transfer by radiation. Consider a black thermocouple in a chamber with black walls.

Suppose the air is at 20°C, the walls are at 100°C, and the convective heat transfer coefficient is $h = 15$ Wm/^{2}K.

What temperature does the thermocouple read?

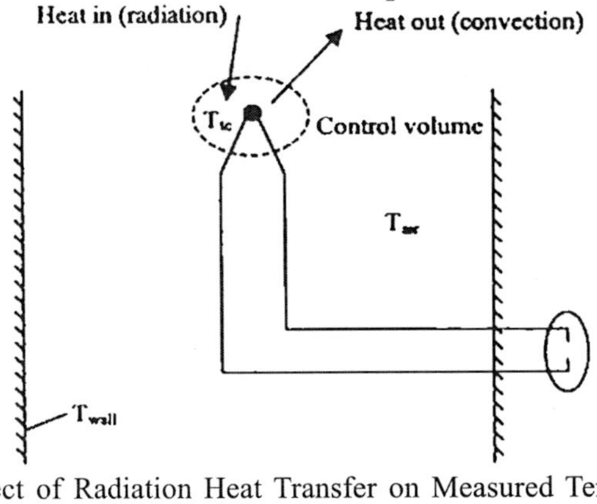

Fig. Effect of Radiation Heat Transfer on Measured Temperature

We use a heat (energy) balance on the control surface. The heat balance states that heat convected away is equal to heat radiated into the thermocouple in steady state. (Conduction heat transfer along the thermocouple wires is neglected here, although it would be included for accurate measurements.)

The heat balance is

$$hA(T_{tc} - T_{air}) = sA(T_{wall}^4 - T_{tc}^4),$$

where A is the area of the thermocouple. Substituting the numerical values gives

$$(15\ \text{Wm}^2\text{-K})T_{tc} - 293\ \text{K}) = (5.67 \times 10^{-8}\ \text{W/m}^2\text{K}^4)\ ((373\ \text{K})^4 - T_{tc}^4)$$

from which we find T_{tc} = 51°C = 324 K. The thermocouple thus sees a higher temperature than the air. We could reduce this error by shielding the thermocouple.

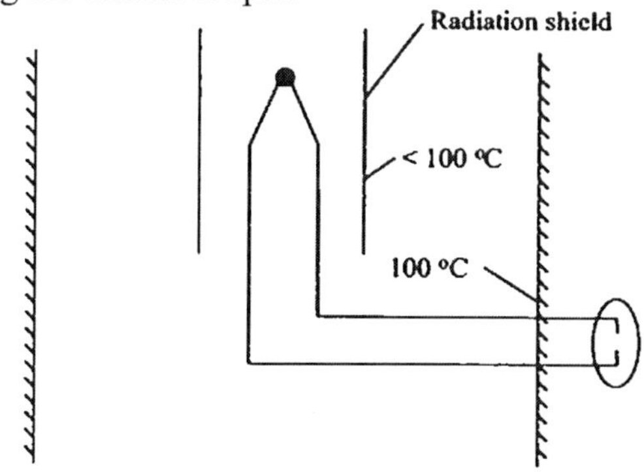

Fig. Shielding a Thermocouple to Reduce Radiation Heat Transfer Error

RADIATION HEAT TRANSFER BETWEEN BLACK SURFACES OF ARBITRARY GEOMETRY

In general, for any two objects in space, a given object 1 radiates to object 2, and to other places as well.

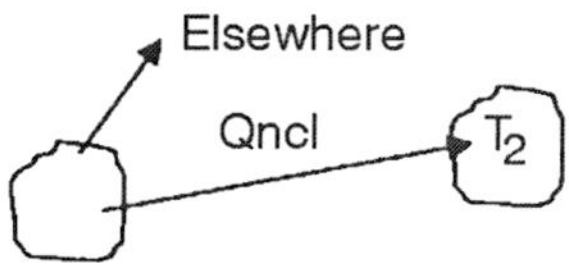

Fig. Radiation between Two Bodies

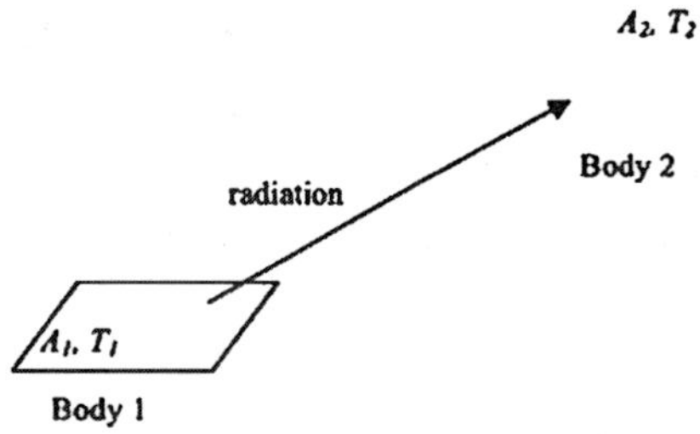

Fig. Radiation between Two Arbitrary Surfaces

We want a general expression for energy interchange between two surfaces at different temperatures. This is given by the radiation shape factor or view factor, F_{i-j}. For the situation in figure,

F_{1-2} = fraction of energy leaving 1 which reaches 2

F_{2-1} = fraction of energy leaving 2 which reaches 1

F_{1-2} , F_{2-1} are functions of geometry only

For body 1, we know that E_b is the emissive power of a black body, so the energy leaving body 1 is $E_{b1}A_1$. The energy leaving body 1 and arriving (and being absorbed) at body 2 is $E_{b1}A_1F_{1-2}$. The energy leaving body 2 and being absorbed at body 1 is $E_{b2}A_2F_{2-1}$. The net energy interchange from body 1 to body 2 is

$$E_{b1}A_1F_{1-2} - E_{b2}A_2F_{2-1} = \dot{Q}_{1-2}.$$

Suppose both surfaces are at the same temperature so there is no net heat exchange. If so,

$$E_{b1}A_1F_{1-2} - E_{b2}A_2F_{2-1} = 0,$$

but also $E_{b1} = E_{b2}$. Thus

$$A_1F_{1-2} = A_2F_{2-1}.$$

Equation is the shape factor reciprocity relation. The net heat exchange between the two surfaces is

$$\dot{Q}_{1-2} = A_1F_{1-2}(E_{b1} - E_{b2}) \quad [or \;\; A_2F_{2-1}(F_{2-1} - Eb2)]$$

Concentric cylinders or concentric spheres

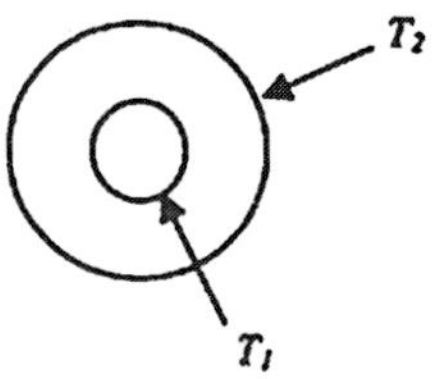

Fig. Radiation Heat Transfer for Concentric Cylinders or Spheres

The net heat transfer from surface 1 to surface 2 of figure is

$$\dot{Q}_{1-2} = A_1 F_{1-2}(E_{b1} - E_{b2})$$

We know that $F_{1-2} = 1$, i.e., that all of the energy emitted by 1 gets to 2. Thus

$$\dot{Q}_{1-2} = A_1(E_{b1} - E_{b2}).$$

This can be used to find the net heat transfer from 2 to 1.

$$\dot{Q}_{1-2} = A_2 F_{2-1}(E_{b2} - E_{b1}) = A_1 F_{1-2}(F_{b2} - E_{b1}) = A_1(E_{b2} - E_{b1}).$$

REFERENCES

- Introduction to Metallurgical Thermodynamics by David R. Gaskell.
- Schaum's Outline of Thermodynamics for Engineers, 2nd edition (Schaum's Outlines) by Merle Potter and Ph.D., Craig Somerton.
- Mixing and Excess Thermodynamic Properties (Physical sciences data) by Jaime Wisniak and Abraham Tamir.
- Thermodynamics by J.P. Holman.
- Schaum Engineering Thermodynamics (Schaum's Outlines) by Merle Potter.

3

Heat Conduction

TEMPERATURE DISTRIBUTIONS IN THE PRESENCE OF HEAT SOURCES

There are a number of situations in which there are sources of heat in the domain of interest. Examples are:

- Electrical heaters where electrical energy is converted resistively into heat.
- Nuclear power supplies.
- Propellants where chemical energy is the source.

These situations can be analyzed by looking at a model problem of a slab with heat sources (W/m^3) distributed throughout. We take the outside walls to be at temperature T_w and we will determine the maximum internal temperature.

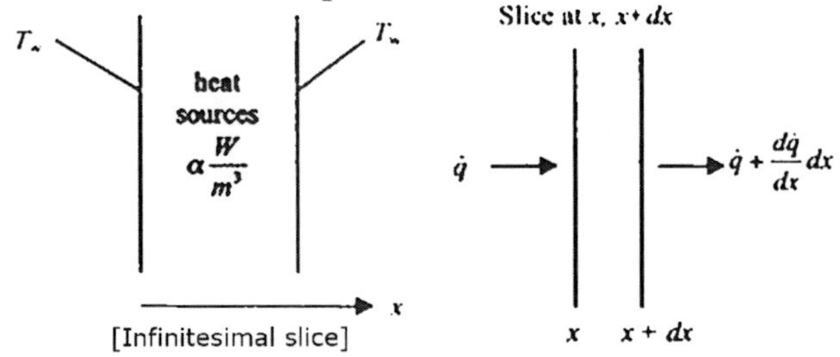

Fig. Slab with Heat Sources

With reference to Figure, a steady-state energy balance yields an equation for the heat flux, :

$$\dot{q} + adx - \left(\dot{q} + \frac{d\dot{q}}{dx} dx \right) = 0.$$

or

$$\frac{d\dot{q}}{dx} = \alpha.$$

There is a change in heat flux with x due to the presence of the heat sources. The equation for the temperature is

$$\frac{d^2T}{dx^2} + \frac{\alpha}{k} = 0.$$

Equation can be integrated once,

$$\frac{dT}{dx} = -\frac{\alpha}{k}x + a,$$

and again to give

$$T = -\frac{\alpha}{2k}x^2 + ax + b,$$

where and are constants of integration. The boundary conditions imposed are $T(0) = T(L) = T_w$. Substituting these into Equation gives $b = T_w$ and $a = \alpha L/2k$. The temperature distribution is thus

$$T = -\frac{\alpha}{2k}x^2 + \frac{\alpha}{2k}Lx + T_w.$$

Writing in a normalized, non-dimensional fashion gives a form that exhibits in a more useful manner the way in which the different parameters enter the problem:

$$\frac{T - T_w}{\alpha L^2 / k} = \frac{1}{2}\left(\frac{x}{L} - \frac{x^2}{L^2}\right).$$

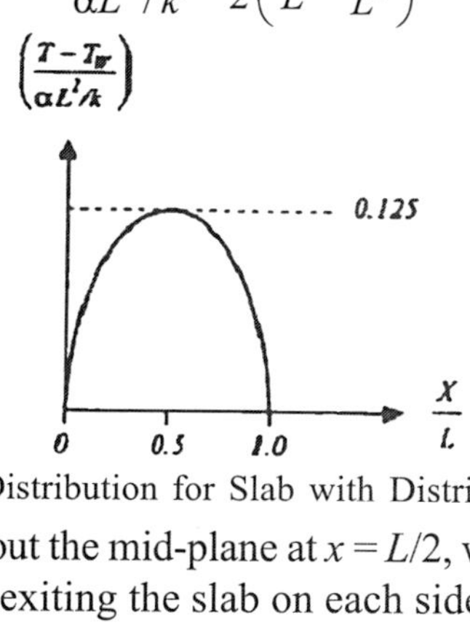

Fig. Temperature Distribution for Slab with Distributed Heat Sources

It is symmetric about the mid-plane at $x = L/2$, with half the energy due to the sources exiting the slab on each side.

The heat flux at the side of the slab, $x = 0$, can be found by differentiating the temperature distribution and evaluating at $x = 0$:

$$-k\left.\frac{dT}{dx}\right|_{x=0} = -k\frac{\alpha L^2}{2k}\frac{1}{L} = -\frac{\alpha L}{2}.$$

This is half of the total heat generated within the slab. The magnitude of the heat flux is the same at $x = L$, although the direction is opposite.

A related problem would be one in which there were heat flux (rather than temperature) boundary conditions at $x = 0$ and $x = L$, so that T_w is not known. We again determine the maximum temperature. At $x = 0$ and L, the heat flux and temperature are continuous so

$$-k\frac{dT}{dx} = h(T - T_\infty) \text{ at } x = 0, \text{L}$$

Referring to the temperature distribution of Equation gives for the two terms in Equation,

$$k\frac{dT}{dx} = k\left(-\frac{\alpha x}{k} + a\right) = (-\alpha x + ka).$$

$$h(T - T_\infty) = h\left(\frac{\alpha x^2}{2k} + ax + b - T_\infty\right).$$

Evaluating Equation at $x = 0$ and L allows determination of the two constants a and b. This is left as an exercise for the reader.

HEAT TRANSFER FROM A FIN

Fins are used in a large number of applications to increase the heat transfer from surfaces. Typically, the fin material has a high thermal conductivity. The fin is exposed to a flowing fluid, which cools or heats it, with the high thermal conductivity allowing increased heat being conducted from the wall through the fin. The design of cooling fins is encountered in many situations and we thus examine heat transfer in a fin as a way of defining some criteria for design.

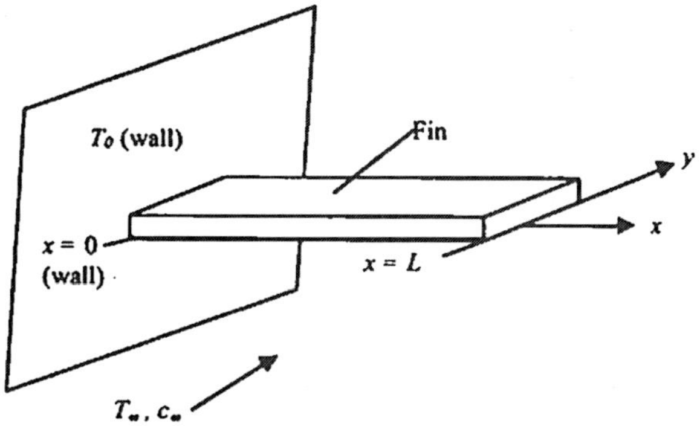

Fig. Geometry of Heat Transfer Fin

The fin is of length L. The other parameters of the problem are indicated. The fluid has velocity $c_\infty$and temperature T_∞. We assume (using the Reynolds analogy or other approach) that the heat transfer coefficient for the fin is known and has the value h. The end of the fin can have a different heat transfer coefficient, which we can call h_L.

The approach taken will be quasi-one-dimensional, in that the temperature in the fin will be assumed to be a function of only. This may seem a drastic simplification, and it needs some explanation. With a fin cross-section equal to A and a perimeter P, the characteristic dimension in the transverse direction is A/P (For a circular fin, $A/P = r/2$). The regime of interest will be taken to be that for which the Biot number is much less than unity, $Bi = h\ (A/P/k \ll 1)$, which is a realistic approximation in practice.

The physical content of this approximation can be seen from the following. Heat transfer per unit area out of the fin to the fluid is roughly of magnitude ~ $h(T_w - T_\infty)$ per unit area. The heat transfer per unit area within the fin in the transverse direction is (again in the same approximate terms)

$$k\frac{(T_1 - T_w)}{A/P},$$

where T_1 is an internal temperature. These two quantities must be of the same magnitude. If $h(A/P)/k \ll 1$, then $(T_1 - T_w)/(T_w - T_\infty)$. In other words, if Bi $\ll 1$, there is a much larger capability for heat transfer per unit area across the fin than there is between the fin and the fluid, and thus little variation in temperature inside the fin in the transverse direction. To emphasize the point, consider the limiting case of zero heat transfer to the fluid, i.e., an insulated fin. Under these conditions, the temperature within the fin would be uniform and equal to the wall temperature.

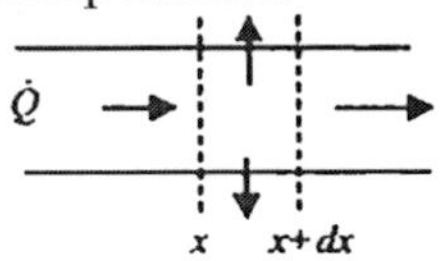

Fig. Element of Fin Showing Heat Transfer

If there is little variation in temperature across the fin, an appropriate model is to say that the temperature within the fin is a function of x only, $T = T(x)$, and use a quasi-one-dimensional approach. To do this, consider an element, dx, of the fin. There is heat flow of magnitude $\dot{Q}_{\text{in}}$ at the left-hand side and heat flow out of magnitude $\dot{Q}_{\text{out}} = \dot{Q}_{\text{in}} + \frac{d\dot{Q}}{dx}dx$ at the right hand side.

There is also heat transfer around the perimeter on the top, bottom, and sides of the fin. From a quasi-one-dimensional point of view, this is a situation similar to that with internal heat sources, but here, for a cooling fin, in each elemental slice of thickness there is essentially a heat sink of magnitude $Pdxh(T - T_\infty)$, where Pdx is the area for heat transfer to the fluid.

The heat balance for the element in Figure can be written in terms of the heat flux using $\dot{Q} = \dot{q}$A, for a fin of constant area:

$$\dot{q}A = Ph(T - T_\infty)dx + \left(\dot{q}A + \frac{d\dot{q}}{dx}dxA\right).$$

From Equation we obtain

$$A\frac{d\dot{q}}{dx} + Ph\,(T - T_\infty) = 0.$$

In terms of the temperature distribution, $T(x)$:

$$\frac{d^2T}{dx^2} - \frac{Ph}{Ak}(T - T_\infty) = 0.$$

The quantity of interest is the temperature difference $(T - T_\infty)$, and we can change variables to put Equation in terms of this quantity using the substitution

$$\frac{d}{dx}(T - T_\infty) = \frac{dT}{dx}.$$

Equation can therefore be written as

$$\frac{d^2}{dx^2}(T - T_\infty) - \frac{Ph}{Ak}(T - T_\infty) = 0.$$

Equation describes the temperature variation along the fin. It is a second order equation and needs two boundary conditions. The first of these is that the temperature at the end of the fin that joins the wall is equal to the wall temperature. (Does this sound plausible? Why or why not?)

$$(T - T_\infty)_{x=0} = T_0 - T_\infty.$$

The second boundary condition is at the other end of the fin. We will assume that the heat transfer from this end is negligible. The boundary condition at $x = L$ is

$$\left.\frac{d}{dx}(T - T_\infty)\right|_{x=L} = 0.$$

The last step is to work in terms of non-dimensional variables to obtain a more compact description. In this we define as $(T - T_\infty)/(T_0 - T_\infty)$, as $\Delta\bar{T}$ where the values of $\Delta\bar{T}$ range from zero to one. We also define $\xi = x/L$, where ξ also ranges over zero to one. The relation between derivatives that is needed to cast the equation in terms of ξ is

$$\frac{d}{dx} = \frac{d\xi}{dx}\frac{d}{d\xi} = \frac{1}{L}\frac{d}{d\xi}.$$

Equation can be written in this dimensionless form as

$$\frac{d^2\Delta\bar{T}}{d\xi^2} - \left(\frac{hP}{kA}L^2\right)\Delta\tilde{T} = 0.$$

There is one non-dimensional parameter in Equation, which we will call m and define by

$$m^2L^2 = \frac{hPL^2}{kA}.$$

The equation for the temperature distribution we have obtained is

$$\frac{d^2\Delta\tilde{T}}{d\xi^2} - m^2L^2\Delta\tilde{T} = 0.$$

This second order equation has the solution

$$\Delta\tilde{T} = ae^{mL\xi} + be^{-mL\xi}.$$

The boundary condition at $\xi = 0$ is

$$\Delta\tilde{T}(0) = a + b = 1.$$

The boundary condition $\xi = 1$ at is that the temperature gradient is zero or

$$\frac{d\Delta\tilde{T}}{d\xi}(L) = mLae^{mL} - mLbe^{-mL} = 0.$$

Solving the two equations given by the boundary conditions for a and b gives an expression for $\Delta\tilde{T}$ in terms of the hyperbolic cosine or :

$$\cosh x = \frac{(e^x + e^{-x})}{2},$$

$$\Delta\tilde{T} = \frac{\cosh mL\,(1-\xi)}{\cosh mL}.$$

This is the solution to Equation for a fin with no heat transfer at the tip. In terms of the actual heat transfer parameters it is written as

$$\frac{T - T_\infty}{T_0 - T_\infty} = \frac{\cosh\left(\left(1 - \frac{x}{L}\right)\sqrt{\frac{hP}{kA}}L\right)}{\cosh\left(\sqrt{\frac{hP}{kA}}L\right)}.$$

The amount of heat removed from the wall due to the fin, which is the quantity of interest, can be found by differentiating the temperature and evaluating the derivative at the wall, :

$$\dot{Q} = -kA\frac{d}{dx}(T - T_\infty)\Big|_{x=0}$$

or

$$\frac{\dot{Q}L}{kA(T_0 - T_\infty)} = -\frac{d\Delta\tilde{T}}{d\xi}\Big|_{x=0} = \frac{mL \sin h(mL)}{\cosh\,(mL)} = mL \tanh\,(mL),$$

$$\frac{\dot{Q}L}{\sqrt{kAhP}(T_0 - T_\infty)} = \tanh\,(mL).$$

The solution is plotted in Figure, which is taken from the book by Lienhard. Several features of the solution should be noted. First, one does not need fins which have a length such that is much greater than 3. Second, the assumption about no heat transfer at the end begins to be inappropriate as gets smaller than 3, so for very short fins the simple expression above would not be a good estimate. We will see below how large the error is.

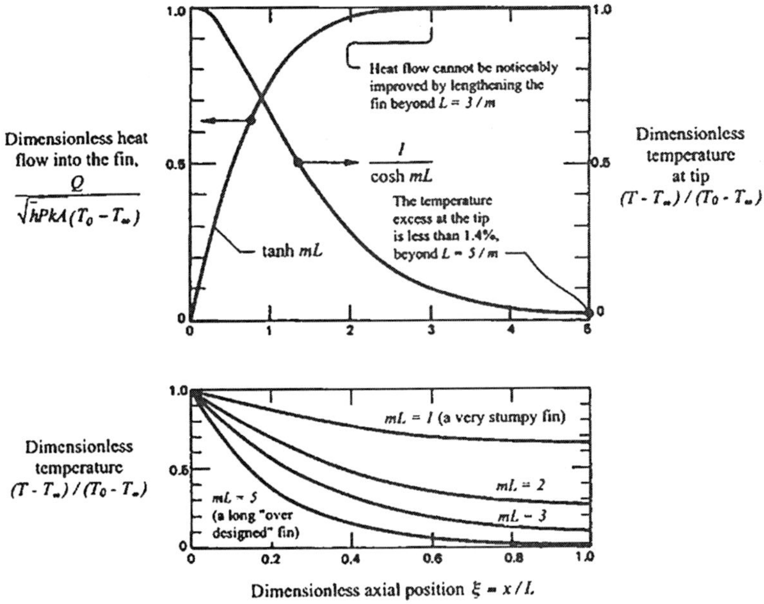

Fig. The Temperature Distribution, Tip Temperature and Heat Flux in a Straight One-dimensional Fin with the Tip Insulated.

TRANSIENT HEAT TRANSFER

All the heat transfer problems we have examined have been steady state, but there are often circumstances in which the transient response to heat transfer is critical. An example is the heating up of gas turbine compressors as they are brought up to speed during take-off.

The disks that hold the blades are large and take a long time to come to temperature, while the casing is thin and in the path of high velocity compressor flow and thus comes to temperature rapidly.

The result is that the case expands away from the blade tips, sometimes enough to cause serious difficulties with aerodynamic performance.

To introduce the topic as well as to increase familiarity with modeling of heat transfer problems, we examine a lumped parameter analysis of an object cooled by a stream. This will allow us to see what the relevant non-dimensional parameters are and, at least in a qualitative fashion, how more complex heat transfer objects will behave. We want to view the object as a "lump" described by a single parameter. We need to determine when this type of analysis would be appropriate. To address this, consider the temperature difference $T_1 - T_w$ between two locations in the object.

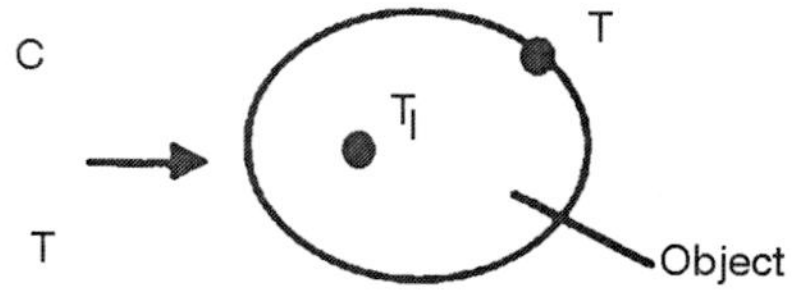

Fig. Temperature Variation in an Object Cooled by a Flowing Fluid

If the heat transfer within the body and from the body to the fluid are of the same magnitude,

$$h(T_w - T_\infty) \approx \frac{k}{L}(T_1 - T_w),$$

where L is a relevant length scale, say half the thickness of the object. The ratio of the temperature difference is

$$\frac{T_1 - T_w}{T_w - T_\infty} \approx \frac{hL}{k}.$$

If the Biot number is small the ratio of temperature differences described in Equation is also $(T_1 - T_w)/(T_w - T_\infty) \ll 1$. We can thus say $(T_1 - T_w) \ll T_w - T_\infty$ and neglect the temperature non-uniformity within the object.

The approximation made is to view the object as having a spatially uniform temperature that is a function of time only. Explicitly, $T = T(t)$. The first law applied to the object is (using the fact that for solids $c_p = c_v = c$),

$$\dot{Q}_{\text{in}} = \rho Vc\frac{dT}{dt},$$

where ρ is the density of the object and is its volume. In terms of heat transferred to the fluid, $\dot{Q}_{\text{out}} = -\rho VcdT/dt$. The rate of heat transfer to the fluid is $Ah(T - T_{\infty}$, so the expression for the time evolution of the temperature is

$$Ah(T - T_{\infty}) = \rho Vc\frac{dT}{dt}.$$

The initial temperature, $T(0)$, is equal to some known value, which we can call T_i. Using this, Equation can be written in terms of a non-dimensional temperature difference $(T - T_{\infty})/(T_i - T_{\infty})$,

$$\frac{d}{dt}\left(\frac{T_0 - T_{\infty}}{T_i - T_{\infty}}\right) + \frac{hA}{\rho Vc}\left(\frac{T - T_{\infty}}{T_i - T_{\infty}}\right) = 0.$$

At time $t = 0$, this non-dimensional quantity is equal to one.

Equation is an equation you have seen before, $\left(\frac{dx}{dt} + \frac{x}{r} = 0\right)$ which has the solution $x = ae^{-t/r}$. For the present problem the form is

$$\frac{T - T_{\infty}}{T_i - T_{\infty}} = ae^{-hAt/\rho Vc}.$$

The constant can be seen to be equal to unity to satisfy the initial condition. This form of equation implies that the solution has a heat transfer "time constant" given by $\tau = \rho Vc/hA$.

The time constant, τ, is in accord with our intuition, or experience; high density, large volume, or high specific heat all tend to increase the time constant, while high heat transfer coefficient and large area will tend to decrease the time constant. This is the same form of equation and the same behaviour you have seen for the R-C circuit, as shown schematically in Figure. The time dependence of the voltage in the R-C circuit when the switch is opened suddenly is given by the equation

$$\frac{dE}{dt} + \frac{E}{RC} = 0.$$

There are, in fact, a number of physical processes which have (or can be modeled as having) this type of exponentially decaying behaviour.

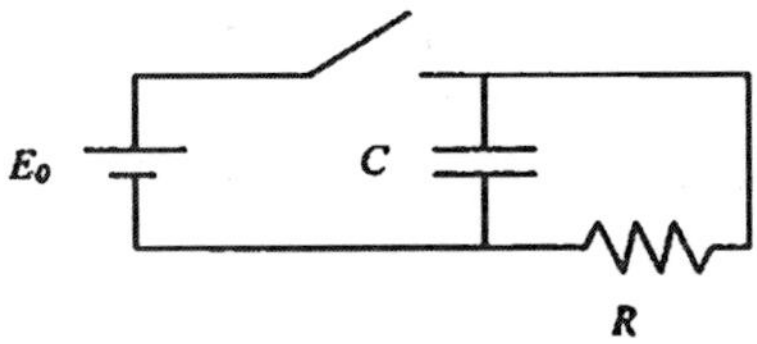

Fig. Voltage Change in an R-C Circuit

MODELING COMPLEX PHYSICAL PROCESSES

A number of assumptions were made about the processes that we were attempting to describe. These are all part of the general approach to modeling of physical systems. The main idea is that for engineering systems, one almost always cannot compute the process exactly, especially for fluid flow problems. At some level of detail, one generally needs to model, i.e. to define some plausible behaviour for attributes of the system that will not be computed.

Modeling can span an enormous range from the level of our assumption of uniform temperature within the solid object to a complex model for the small scale turbulent eddies in the flow past a compressor blade.

In carrying out such modeling, it is critical to have a clear idea of just what the assumptions really mean, as well as the fidelity that we ascribe to the descriptions of actual physical phenomena, and we thus look at the statements we have made in this context.

One assertion made was that because $hL/k \ll 1$ and on the basis of a heat balance,

$$\frac{k}{L}(T_c - T_w) \approx h(T_w - T_\infty),$$

we could assume

$$\frac{(T_{\text{body interior}} - T_w)}{(T_w - T_\infty)} \ll 1.$$

Based on this, we said that T_{body} is approximately uniform and $T_w \approx T_{\text{body interior}}$. Another aspect is that setting erases any geometrical detail of the fin cross section. The only place where P and A enter the problem is in a non-dimensional combination. A third assumption, made in the fin problem, was that the heat transfer at the far end can be neglected. The solution including this effect, where $h_L L/k$ is an axial Biot number is given as Equation. If the quantity $h_L L/k$ is small, you can see that Equation reduces to the previous result.

$$\frac{T - T_\infty}{T_0 - T_\infty} = \frac{\cosh\,(mL(1-\xi)) + (Bi_{\text{axial}}/m)\sin\text{h}(mL(1-\xi))}{\cosh\,(mL) + (Bi_{\text{axial}}/mL)\,\sin\;h\,(mL)}$$

and

$$\frac{\dot{Q}}{\sqrt{kAhP}(T_0 - T_\infty)} = \frac{Bi_{\text{axial}}/mL + \tan\,h\,(mL)}{1 + \dfrac{Bi_{\text{axial}}}{mL}\tan\,h\,(mL)}.$$

HEAT EXCHANGERS

The general function of a heat exchanger is to transfer heat from one fluid to another. The basic component of a heat exchanger can be viewed as a tube with one fluid running through it and another fluid flowing by on the outside. There are thus three heat transfer operations that need to be described:

- Convective heat transfer from fluid to the inner wall of the tube,
- Conductive heat transfer through the tube wall, and
- Convective heat transfer from the outer tube wall to the outside fluid.

Heat exchangers are typically classified according to flow arrangement and type of construction. The simplest heat exchanger is one for which the hot and cold fluids move in the same or opposite directions in a concentric tube (or double-pipe) construction.

In the parallel-flow arrangement of Figure, the hot and cold fluids enter at the same end, flow in the same direction, and leave at the same end. In the counterflow arrangement of Figure, the fluids enter at opposite ends, flow in opposite directions, and leave at opposite ends.

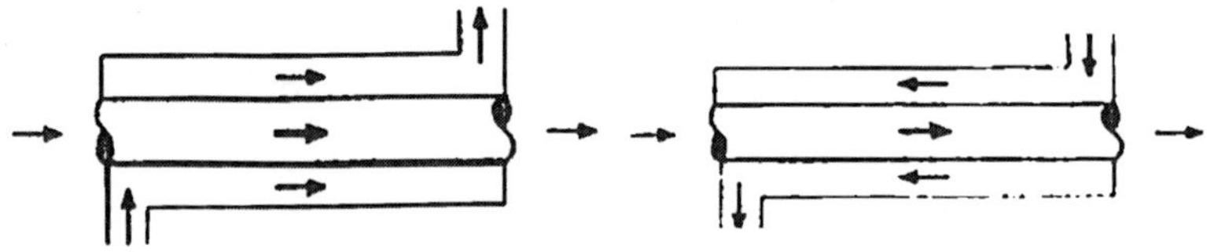

[Parallel flow] [Counter flow]

Fig. Concentric Tubes heat Exchangers

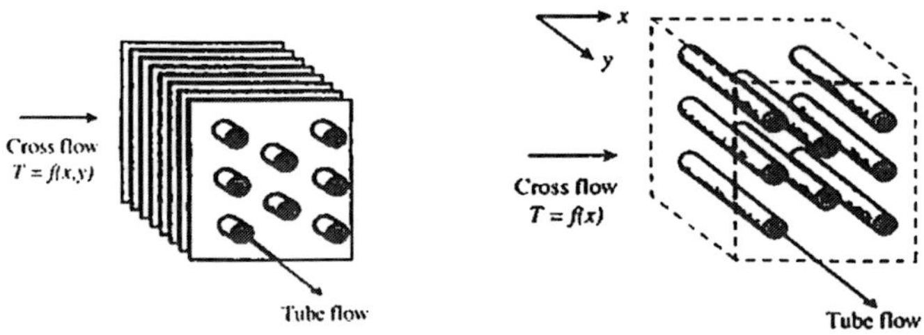

[Finned with both fluids unmixed] [Unfinned with one fluid mixed and the other unmixed]

Fig. Cross-flow Heat Exchangers

Alternatively, the fluids may be in cross flow (perpendicular to each other), as shown by the finned and unfinned tubular heat exchangers of Figure.

The two configurations differ according to whether the fluid moving over the tubes is unmixed or mixed. The fluid is said to be unmixed because the fins prevent motion in a direction (x) that is transverse to the main flow direction (y).

In this case the fluid temperature varies with x and y. In contrast, for the unfinned tube bundle of Figure, fluid motion, hence mixing, in the transverse direction is possible, and temperature variations are primarily in the main flow direction. Since the tube flow is unmixed, both fluids are unmixed in the finned exchanger, while one fluid is mixed and the other unmixed in the unfinned exchanger.

To develop the methodology for heat exchanger analysis and design, we look at the problem of heat transfer from a fluid inside a tube to another fluid outside.

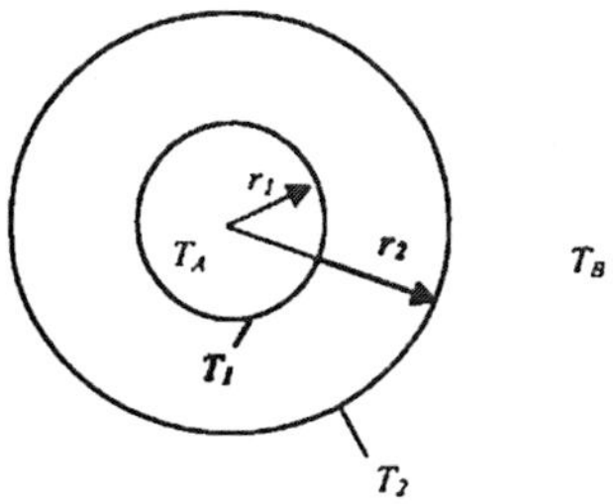

Fig. Geometry for Heat Transfer between Two Fluids

We examined this problem before in Section and found that the heat transfer rate per unit length is given by

$$\dot{Q} = \frac{2pk(T_A - T_B)}{\dfrac{k}{r_1h_1} + \dfrac{k}{r_2h_2} + \ln\left(\dfrac{r_2}{r_1}\right)}.$$

Here we have taken into account one additional thermal resistance, the resistance due to convection on the interior, and include in our expression for heat transfer the bulk temperature of the fluid, T_A, rather than the interior wall temperature, T_1.

It is useful to define an overall heat transfer coefficient h_0 per unit length as

$$\dot{Q} = 2\pi r_2 h_0 (T_A - T_B)$$

From Equations the overall heat transfer coefficient, h_0, is

$$\frac{1}{h_0} = \frac{r_2}{r_1h_1} + \frac{r_2}{k}\ln\left(\frac{r_2}{r_1}\right) + \frac{1}{h_2}.$$

We will make use of this in what follows.

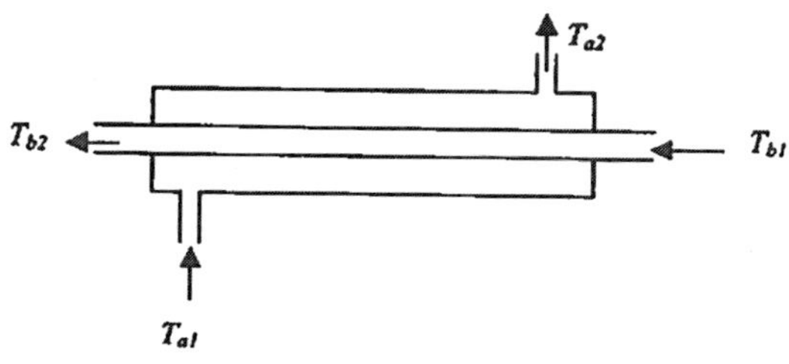

Fig. Counterflow Heat Exchanger

A schematic of a counterflow heat exchanger. We wish to know the temperature distribution along the tube and the amount of heat transferred.

SIMPLIFIED COUNTERFLOW HEAT EXCHANGER

To address this we start by considering the general case of axial variation of temperature in a tube with wall at uniform temperature T_0 and a fluid flowing inside the tube.

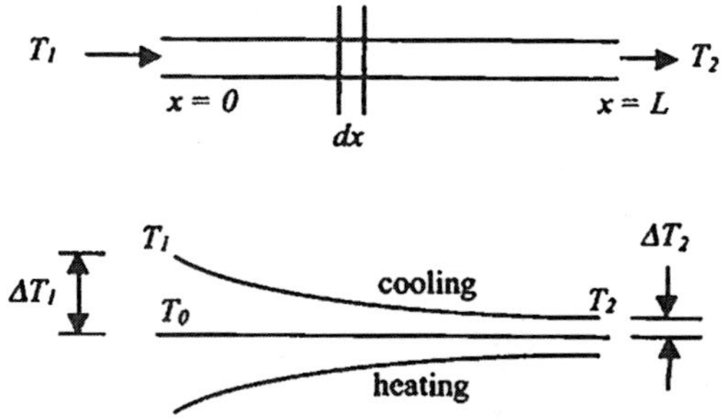

Fig. Fluid Temperature Distribution Along the Tube with Uniform Wall Temperature

The objective is to find the mean temperature of the fluid at x, $T(x)$, in the case where fluid comes in at $x = 0$ with temperature T_1 and leaves at $x = L$ with temperature T_2. The expected distribution for heating and cooling are sketched in Figure.

For heating ($T_0 > T$), the heat flow from the pipe wall in a length dx is

$$\dot{q}\pi\Delta dx = h\pi D(T_0 - T)dx,$$

where D is the pipe diameter. The heat given to the fluid (the change in enthalpy) is given by

$$\rho u_m c_p \frac{\pi D^2}{4} dT = \dot{m} c_p dT,$$

where ρ is the density of the fluid, u_m is the mean velocity of the fluid, c_p is the specific heat of the fluid and $\dot{m}$ is the mass flow rate of the fluid. Setting the last two expressions equal and integrating from the start of the pipe, we find

$$\int_{T_1}^{T} \frac{dT}{T_0 - T} = \int_0^x \frac{4h}{\rho u_m c_p D} dx.$$

Carrying out the integration,

$$\frac{4hx}{\rho u_m c_p D} = \int_{T_1}^{T} \frac{dT}{T_0 - T} = -\int_{T_1}^{T} \frac{dT}{T_0 - T} = -\int_{T_1}^{T} \frac{d(T_0 - T)}{T_0 - T} = -\ln(T_0 - T)\Big|_{T_1}^{T},$$

i.e.,

$$\ln\left(\frac{T_0 - T}{T_0 - T_1}\right) = -\frac{4hx}{\pi u_m c_p D}.$$

Equation can be written as

$$\frac{T_0 - T}{T_0 - T_1} = e^{-\beta x},$$

where

$$\beta = \frac{4h}{\rho u_m c_p D} = \frac{\pi h D}{\dot{m} c_p}.$$

This is the temperature distribution along the pipe. The exit temperature at $x = L$ is

$$\frac{T_0 - T_2}{T_0 - T_1} = e^{\frac{-\pi h D L}{\dot{m} c_p}}.$$

The total heat transfer to the wall all along the pipe is

$$\dot{Q} = \dot{m} c_p (T_1 - T_2).$$

From Equation,

$$\dot{m} c_p = \frac{h\pi D L}{\left(\frac{T_0 - T_1}{T_0 - T_2}\right)}.$$

The total rate of heat transfer is therefore

$$\dot{Q} = \frac{h\pi D L (T_1 - T_2)}{\ln\left(\frac{T_1 - T_0}{T_2 - T_0}\right)}$$

or

$$\dot{Q} = h\pi D L \Delta T_{LM},$$

where ΔT_{LM} is the logarithmic mean temperature difference, defined as

$$DT_{LM} = \frac{T_2 - T_1}{\ln\left(\frac{T_0 - T_2}{T_0 - T_1}\right)} = \frac{\Delta T_1 - \Delta T_2}{\ln\left(\frac{\Delta T_1}{DT_2}\right)}.$$

The concept of a logarithmic mean temperature difference is useful in the analysis of heat exchangers. We will define a logarithmic mean temperature difference for the general counterflow heat exchanger below.

GENERAL COUNTERFLOW HEAT EXCHANGER

We return to our original problem, and write an overall heat balance between the two counterflowing streams as

$$\dot{Q} = \dot{m}_a c_{pa}(T_{a1} - T_{a2}) = \dot{m}_b c_{pb}(T_{b2} - T_{b1}).$$

From a local heat balance, the heat given up by stream in length d_x is $-\dot{m}_a c_{pa} dT_a$. (There is a negative sign since T_a decreases). The heat taken up by stream b is $-\dot{m}_b c_{pa} dT_b$. (There is a negative sign because T_b decreases as x increases). The local heat balance is

$$-\dot{m}_a c_{pa} dT_b = -\dot{m}_b c_{pb} dT_b = \dot{q}dA = \dot{q}\pi Dx.$$

Solving for dT_a and dT_b, we find

$$dT_a = -\frac{\dot{q}dA}{\dot{m}_a c_{pa}};\ dT_b = \frac{\dot{q}dA}{\dot{m}_a c_{pb}}.$$

$$d(T_a - T_b) = d\Delta T = -\left(\frac{1}{\dot{m}_a c_{pa}} - \frac{1}{\dot{m}_b c_{pb}}\right)\dot{q}dA = -\left(\frac{1}{W_a} - \frac{1}{W_b}\right)\dot{q}\pi Ddx,$$

where $W = \dot{m}c_p$. Also, $\dot{q} = h_0 \Delta T$ where h_0 is the overall heat transfer coefficient. We can then say

$$\frac{d\Delta T}{\Delta T} = -h_0 \pi \Delta\left(\frac{1}{W_a} - \frac{1}{W_b}\right)dx.$$

Integrating from $x = 0$ to x gives

$$\ln\left(\frac{T_{a2} - T_{b1}}{T_{a1} - T_{b2}}\right) = -h_0 \pi DL\left(\frac{1}{W_a} - \frac{1}{W_b}\right).$$

Equation can also be written as

$$\frac{T_{a2}-T_{b1}}{T_{a1}-T_{b2}}=e^{-a},$$

where

$$\alpha=h_0\pi DL\left(\frac{1}{W_a}-\frac{1}{W_b}\right).$$

We know that

$$T_{a1}-T_{a2}=\frac{\dot{Q}}{W_a};\ T_{b2}-T_{b1}=\frac{\dot{Q}}{W_b}.$$

Thus

$$(T_{a1}-T_{b2})-(T_{a2}-T_{b1})=\dot{Q}\left(\frac{1}{W_a}-\frac{1}{W_b}\right).$$

Solving for the total heat transfer:

$$\dot{Q}=\frac{(T_{a1}-T_{b2})-(T_{a2}-T_{b1})}{\left(\frac{1}{W_a}-\frac{1}{W_b}\right)}.$$

Rearranging allows us to express $\left(\frac{1}{W_a}-\frac{1}{W_b}\right)$ in terms of other parameters as

$$\left(\frac{1}{W_a}-\frac{1}{W_b}\right)=-\frac{\ln\left(\frac{T_{a2}-T_{b1}}{T_{a1}-T_{b2}}\right)}{h_0\pi DL}.$$

Substituting Equations we obtain a final expression for the total heat transfer for a counterflow heat exchanger:

$$\dot{Q}=h_0\pi DL\frac{(T_{a1}-T_{b2})-(T_{a2}-T_{b1})}{\ln\left(\frac{T_{a1}-T_{b2}}{T_{a2}-T_{b1}}\right)}$$

or

$$\dot{Q}=h_0\pi DL\Delta T_{LM}.$$

EFFICIENCY OF A COUNTERFLOW HEAT EXCHANGER

Suppose we know only the two inlet temperatures T_{a1}, T_{b1}, and we need to find the outlet temperatures. From,

$$T_{a2} - T_{b1} = (T_{a1} - T_{b2})\, e^{-\alpha}$$

or, rearranging,

$$T_{a2} - T_{a1} = T_{b1} - T_{a1} + (T_{a1} - T_{b2})\, e^{-\alpha}.$$

Eliminating $\dot{Q}$ from Equation,

$$T_{b2} = T_{b1} + \frac{W_a}{W_b}(T_{a1} - T_{a2}).$$

We now have two equations, Equation and two unknowns, T_{a2} and T_{b2}. Solving first for T_{a2},

$$T_{a2} - T_{a1} = T_{b1} - T_{a1} + (T_{a1} - T_{b1})\, e^{-\alpha} \frac{W_a}{W_b}(T_{a1} - T_{a2})\, e^{-\alpha}$$

or

$$(T_{a1} - T_{b2})\left(1 - \frac{W_a}{W_b} e^{-\alpha}\right) = (T_{a1} - T_{a2})\,(1 - e^{-\alpha}).$$

$$(T_{a1} - T_{b2}) = \eta\,(T_{a1} - T_{a1}),$$

where η is the efficiency of a counterflow heat exchanger:

$$\eta = \frac{1 - e^{-\alpha}}{1 - \frac{W_a}{W_b} e^{-\alpha}} = \frac{1 - e^{-\alpha}}{1 - \frac{\dot{m}_a c_{pa}}{\dot{m}_b c_{pa}} e^{-\alpha}}.$$

Equation gives T_{a2} in terms of known quantities. We can use this result in to find T_{b2}:

$$T_{b2} - T_{b1} = \frac{\dot{m}_a c_{pa}}{\dot{m}_b c_{pb}}(T_{a1} - T_{a2}) = \frac{\dot{m}_a c_{pa}}{\dot{m}_b c_{pb}}\,\eta\,(T_{a1} - T_{b1}).$$

We examine three examples.

- $\dot{m}_b c_{pb} > \dot{m}_a c_{pa}$

  DT can approach zero at cold end.

  $$\pi D L h_0 \left(\frac{1}{\dot{m}_a c_{pa}} - \frac{1}{\dot{m}_b c_{pb}}\right).$$

h → 1as h_0, surface area, .

Maximum value of ratio $\frac{T_{a1} - T_{a2}}{T_{a1} - T_{b1}} = 1$

Maximum value of ratio $\frac{T_{b2} - T_{b2}}{T_{a1} - T_{b1}} = \frac{\dot{m}_a c_{pa}}{\dot{m}_b c_{pb}}$.

- $\dot{m}_b c_{pb} < \dot{m}_a c_{pa}$

a is negative, $\eta \rightarrow \frac{\dot{m}_b c_{pb}}{\dot{m}_a c_{pa}}$ as [ ] ® ¥ $(W_b < W_a)$

Maximum value of ratio $\frac{T_{a2} - T_{a2}}{T_{a1} - T_{b1}} = \frac{\dot{m}_b c_{pb}}{\dot{m}_a c_{pa}}$

Maximum value of ratio $\frac{T_{a2} - T_{a1}}{T_{a1} - T_{b1}} = 1$.

- $\dot{m}_b c_{pb} < \dot{m}_a c_{pa}$

$$d\,(T_a - T_b) = 0$$

temperature difference remains uniform, $\eta = 1$.

References

- DK Science Encyclopedia by Susan McKeever and Martyn Foote.
- Course in Thermodynamics . Revised Printing. Volume II. (Series in Thermal and Fluids Engineering) by Joseph Kestin.
- Essentials of Thermodynamics (Essentials) by Research & Education Association, Rea, and Staff of Research Education Association .
- Introductory Statistical Mechanics, R. Bowley and M. Sánchez
- Equilibrium Statistical Physics, M. Plische and B. Bergersen

4

The Behaviour of Gases

IDEAL GAS MODEL, HEAT, WORK AND HERMODYNAMICS

The Kinetic Theory picture of a gas (outlined in the previous lecture) is often called the *Ideal Gas Model*. It ignores interactions between molecules, and the finite size of molecules. In fact, though, these only become important when the gas is very close to the temperature at which it become liquid, or under extremely high pressure.

In this lecture, we will be analyzing the behaviour of gases in the pressure and temperature range corresponding to heat engines, and in this range the Ideal Gas Model is an excellent approximation. Essentially, our Programme here is to learn how gases absorb heat and turn it into work, and *vice versa*. This heat-work interplay is called *thermodynamics*.

Julius Robert Mayer was the first to appreciate that there *is* an equivalence between heat and mechanical work. The tortuous path that led him to this conclusion is described in an earlier lecture, but once he was there, he realized that in fact the numerical equivalence—how many Joules in one calorie in present day terminology—could be figured out easily from the results of some measurements of gas specific heat by French scientists. The key was that they had measured specific heats both at constant *volume* and at constant *pressure*.

Mayer realized that in the latter case, heating the gas necessarily increased its volume, and the gas therefore did work in pushing to expand its container.

Having convinced himself that mechanical work and heat were equivalent, evidently the extra heat needed to raise the temperature of the gas at constant pressure was exactly the work the gas did on its container. The simplest way to see what's going on is to imagine the gas in a cylinder, held in by a piston, carrying a fixed weight, able to move up and down the cylinder smoothly with negligible friction. The pressure on the gas is just the total weight pressing down divided by the area of the piston, and this total weight, of course, will not change as the piston moves slowly up or down: the gas is at *constant pressure*.

GAS

MOVEABLE PISTON

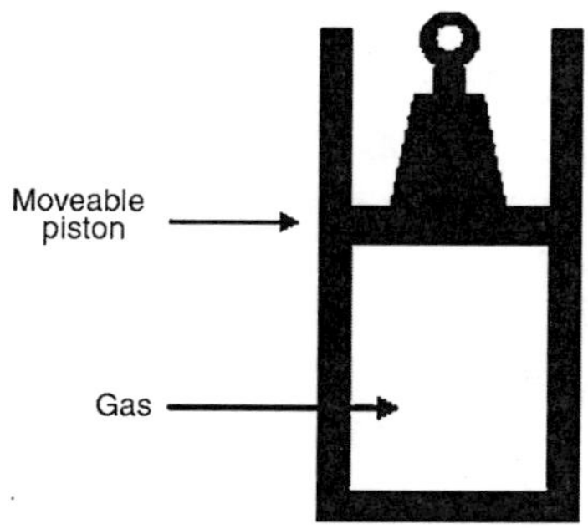

The Gas Specific Heats C_V and C_P

Consider now the two specific heats of this same sample of gas, let's say one mole:

- Specific heat at constant volume, C_V (piston glued in place),
- Specific heat at constant pressure, C_P (piston free to rise, no friction).

In fact, we already worked out C_V in the Kinetic Theory lecture: at temperature T, recall the average kinetic energy per molecule is $\frac{3}{2}kT$, so one mole of gas—Avogadro's number of molecules—will have total kinetic energy, which we'll label *internal energy*,

$$E_{\text{int}} \tfrac{3}{2} kT.N_A = \tfrac{3}{2} RT.$$

That the internal energy is $\frac{3}{2}RT$ per mole immediately gives us the specific heat of a mole of gas in a fixed volume,

$$C_V = \frac{3}{2}R$$

that being the heat which must be supplied to raise the temperature by one degree.

However, if the gas, instead of being in a fixed box, is held in a cylinder at constant pressure, experiment confirms that *more* heat must be supplied to raise the gas temperature by one degree.

As Mayer realized, the total heat energy that must be supplied to raise the temperature of the gas one degree *at constant pressure* is $\frac{3}{2}k$ per molecule *plus* the energy required to lift the weight. The work the gas must do to raise the weight is the force the gas exerts on the piston multiplied by the distance the piston moves. If the area of piston is A, then the gas at pressure P exerts force PA. If on heating through one degree the piston rises a distance Δh, the gas does work

$$PA.\Delta h = P\Delta V.$$

Now, for one mole of gas, $PV = RT$, so at constant P

$$P\Delta V = R\Delta T.$$

Therefore, the work done by the gas in raising the weight is just $R\Delta T$, the specific heat at constant pressure, the total heat energy needed to raise the temperature of one mole by one degree,

$$C_P = C_V + R.$$

In fact, this relationship is true whether or not the molecules have rotational or vibrational internal energy. (It's known as Mayer's relationship.) The specific heat of oxygen at constant volume

$$C_V(O_2) = \frac{5}{2}R$$

and this is understood as a contribution of $\frac{3}{2}R$ from kinetic energy, and R from the two rotational modes of a dumbbell molecule (just why there is no contribution form rotation about the third axis can only be understood using quantum mechanics). The specific heat of oxygen at constant pressure

$$C_P(O_2) = \frac{7}{2}R.$$

It's worth having a standard symbol for the *ratio* of the specific heats:

$$\frac{C_P}{C_V} = \gamma.$$

Tracking a Gas in the (*P*, *V*) Plane: Isotherms and Adiabats

An ideal gas in a box has three thermodynamic variables: P, V, T. But if there is a fixed mass of gas, fixing two of these variables fixes the third from $PV = nRT$ (for n moles). In a heat engine, heat can enter the gas, then leave at a different stage. The gas can expand doing work, or contract as work is done on it. To track what's going on as a gas engine transfers heat to work, say, we must follow the varying state of the gas. We do that by tracing a curve in the (P, V) plane.

Supplying heat to a gas which consequently expands and does mechanical work is the key to the heat engine. But just knowing that a gas is expanding and doing work is not enough information to follow its path in the (P, V) plane. The route it follows will depend on whether or not heat is being supplied (or taken away) at the same time.

There are, however, two particular ways a gas can expand reversibly—meaning that a tiny change in the external conditions would be sufficient for the gas to retrace its path in the (P, V) plane backwards. It's important to concentrate on reversible paths, because as Carnot proved they correspond to the most efficient engines. The two sets of reversible paths are the *isotherms* and the *adiabats*.

Isothermal behaviour: The gas is kept *at constant temperature* by allowing heat flow back and forth with a very large object (a "heat reservoir") at temperature T. From $PV = nRT$, it is evident that for a fixed mass of gas, held at constant T but subject to (slowly) varying pressure, the variables P, V will trace a hyperbolic path in the (P, V) plane.

This path, $PV = nRT_1$, say is called the *isotherm* at temperature T_1. Here are two examples of isotherms:

Adiabatic behaviour: "Adiabatic" means "nothing gets through", in this case *no heat gets in or out* of the gas through the walls. So all the work done in compressing the gas has to go into the internal energy E_{int}.

Adiabatic Compression

Cylinder of insulating material:

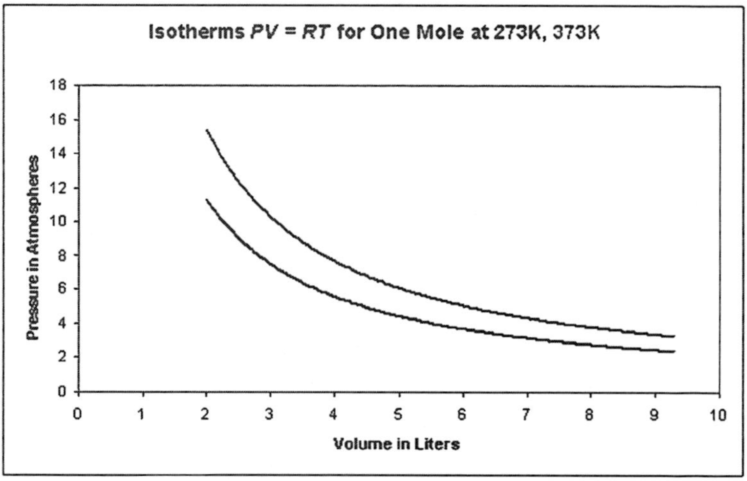

As the gas is compressed, it follows a curve in the (P, V) plane called an *adiabat.*

To see how an adiabat differs from an isotherm, imagine beginning at some point on the blue 273K isotherm on the above graph, and applying pressure so the gas moves to higher pressure and lower volume.

Since the gas's internal energy is increasing, but the number of molecules is staying the same, its temperature is necessarily rising, it will move towards the red curve, then above it. This means *the adiabats are always steeper than the isotherms.*

In the diagram below, we've added a couple of adiabats to the isotherms:

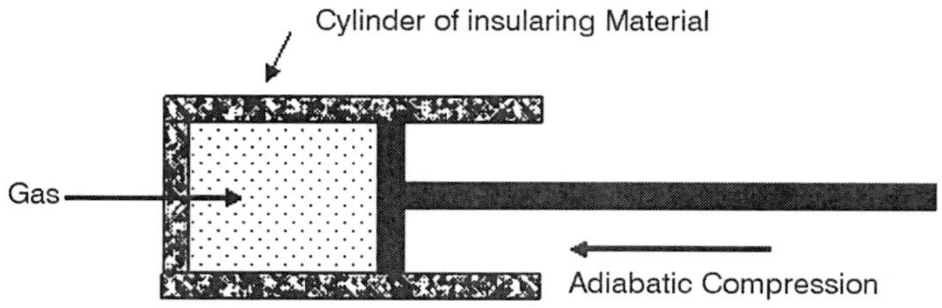

Equation for an Adiabat

What equation for an adiabat corresponds to $PV = nRT_1$ for an isotherm?

On raising the gas temperature by ΔT, the change in the *internal energy*—the sum of molecular kinetic energy, rotational energy and vibrational energy (if any),

$$\Delta E_{\text{int}} = C_V \Delta \boldsymbol{T}.$$

This is always true: whether or not the gas is changing volume is irrelevant, all that counts in E_{int} is the sum of the energies of the individual molecules (assuming as we do here that attractive or repulsive forces between molecules are negligible).

In adiabatic compression, all the work done by the external pressure goes into this internal energy, so

$$-P\Delta V = C_V \Delta T.$$

(Compressing the gas of course gives negative ΔV positive ΔE_{int}.)

To find the equation of an adiabat, we take the infinitesimal limit

$$-PdV = C_V dT$$

Divide the left-hand side by PV, the right-hand side by RT to find

$$-\frac{R}{C_V}\frac{dV}{V} = \frac{dT}{T}.$$

Recall now that $C_P = C_V + R$ and $C_P/C_V = \gamma$ It follows that

$$\frac{R}{C_V} = \frac{C_P - C_V}{C_V} = \gamma - 1.$$

Hence

$$-(\gamma - 1)\int \frac{dV}{V} = \int \frac{dT}{T}$$

and integrating

$$\ln T + (\gamma - 1) \ln V = \text{const.}$$

from which the equation of an adiabat is

$$TV^{\gamma-1} = \text{const}$$

From $PV = RT$, the P, V equation for an adiabat can be found by multiplying the left-hand side of this equation by the constant PV/T, giving

$PV^{\gamma} =$ for an adiabat,

where $\gamma = 5/3$ for a monatomic gas, $1/5$ for a diatomic gas.

REAL GASES VS. IDEAL GASES

Most of the discussions of gases assume that the gases exhibit ideal behaviour. Ideal behaviour involves two things: the first is that the gas can be infinitely compressed or infinitely cooled and the gas will not liquefy. The second is that the gas molecules have no volume. With these assumptions, the ideal gas law, PV=nRT, can be used.

In reality, however, if a gas is compressed enough the particles will attract and will liquefy. Similarly if the gas is cooled to its boiling point, it will liquefy. Therefore at low temperatures or high pressures, the effect of the attractive forces becomes larger. However, if the gas is moving fast enough, attractive forces between the molecules that cause liquefaction are not a factor.

Gas molecules also definitely have a volume, small though it may be, and the volume of the molecules play a factor under conditions of large gas molecules and small container volumes.

Joseph van der Waals studied the behaviour of real gases and made comparisons to the ideal gas law. He derived an equation to account for the differences. The equation adds in two constants, a and b, to the ideal gas law. These constants are derived to give the best agreement between the observed behaviour and the equation. Therefore each gas has its own values for the constants. The van der Waals equation is stated as:

$$\left(P + \frac{n^2 a}{V^2}\right)(V - nb) = nRT$$

$P + n^2a/V^2$ deals with attractive forces between molecules and how they reduce the ideal pressure. $V - nb$ accounts for volume of the particles, where the constant b is related to the size of the molecule and since the molecules take up space, the effective size of the container is decreased.

Van der Waals received a Nobel Prize in physics in 1910 for his work in gases and liquids. Below is a table giving the a and b constants for various gases. The a values are small for those gases with small intermolecular attractions, such as He. In general the larger molecules have a larger b constant, as can be seen for octane, though this is not the only factor for determining *b*.

Gas	*Formula*	*a [(L² · atm)/mole²]*	*b [L/mole]*
Helium	He	0.03412	0.02370
Hydrogen	H_2	0.2444	0.02661
Nitrogen	N_2	1.390	0.03913
Oxygen	O_2	1.360	0.03183
Carbon dioxide	CO_2	3.592	0.04267
Acetylene	C_2H_2	4.390	0.05136
Chlorine	Cl_2	6.493	0.05622
n - Butane	C_4H_{10}	14.47	0.1226
n - Octane	C_8H_{18}	37.32	0.2368

To illustrate the differences between the two equations, an example using acetylene and helium gas will be shown.

Example: One mole of acetylene gas is placed in a 20.0 L container at 25 °C.

- The pressure using the Ideal Gas Law is shown to be:

$$P = \frac{nRT}{V} = \frac{(1\text{ mol})(0.0821\frac{\text{Lat m}}{\text{mol K}})(298\ K)}{200\ L} = 1.223\text{ atm}$$

- The pressure using the van der Waals equation is shown to be:

$$P = \left(\frac{nRT}{V - nb}\right) - \left(\frac{n^2 a}{V^2}\right)$$

$$= \left(\frac{(1\text{ mol})(0.0821)\frac{\text{Latm}}{\text{mol}})(298\text{ K})}{20.0\text{ L} - (1\text{ mol})(0.05136\frac{\text{L}}{\text{mol}})}\right) - \left(\frac{(1\text{ mol})^2\ (4.390\frac{\text{Latm}}{(20.0\text{ L})^2}}{(20.0\text{ L})^2}\right)$$

$$= 1.215\text{ atm}$$

There is approximately a 0.66% difference between the two values. If the same calcul

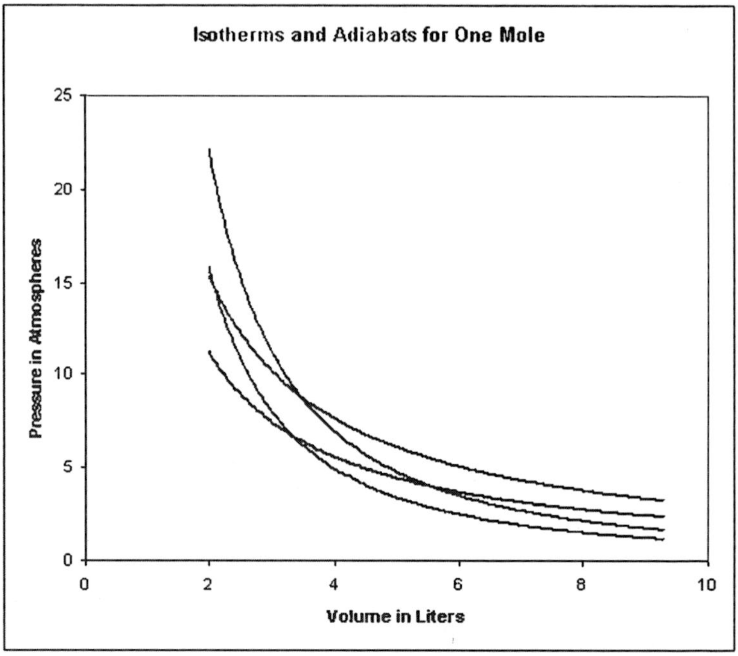

ation was done with helium gas, the difference would only be about 0.13%.

REFERENCE

- Equilibrium Thermodynamics, C. J. Adkins.
- Thermodynamics (and Introduction to Thermostatistics), H. B. Callen.
- Statistical mechanics, R. P. Feynman, W. A. Benjamin.
- An introduction to Statistical Thermodynamics, T.L. Hill
- Statistical Mechanics, K. Huang.

5

Specific Heat of Solids

Consider a simple solid containing atoms. Now, atoms in solids cannot translate (unlike those in gases), but are free to vibrate about their equilibrium positions. Such vibrations are called *lattice vibrations*, and can be thought of as sound waves propagating through the crystal lattice. Each atom is specified by three independent position coordinates, and three conjugate momentum coordinates. Let us only consider small amplitude vibrations. In this case, we can expand the potential energy of interaction between the atoms to give an expression which is quadratic in the atomic displacements from their equilibrium positions.

It is always possible to perform a *normal mode analysis* of the oscillations. In effect, we can find $3N$ independent modes of oscillation of the solid. Each mode has its own particular oscillation frequency, and its own particular pattern of atomic displacements. Any general oscillation can be written as a linear combination of these *normal modes*.

Let q_i be the (appropriately normalized) amplitude of the ith normal mode, and p_i the momentum conjugate to this coordinate. In *normal mode coordinates*, the total energy of the lattice vibrations takes the particularly simple form

$$E = \frac{1}{2}\sum_{i=1}^{3N}(p_i^2 + w_i^2 q_i^2),$$

where w_i is the (angular) oscillation frequency of the th normal mode. It is clear that in normal mode coordinates, the linearized lattice vibrations are equivalent to $3N$ independent harmonic oscillators (of course, each oscillator corresponds to a different normal mode).

The typical value of w_i is the (angular) frequency of a sound wave propagating through the lattice. Sound wave frequencies are far lower than the typical vibration frequencies of gaseous molecules. In the latter case, the mass involved in the vibration is simply that of the molecule, whereas in the former case the mass involved is that of very many atoms (since lattice vibrations are non-localized).

The strength of interatomic bonds in gaseous molecules is similar to those in solids, so we can use the estimate $w \sim \sqrt{k/m}$ (k is the force constant which measures the strength of interatomic bonds, and is the mass involved in the oscillation) as proof that the typical frequencies of lattice vibrations are very much less than the vibration frequencies of simple molecules. It follows from $\Delta E = \hbar w$ that the quantum energy levels of lattice vibrations are far more closely spaced than the vibrational energy levels of gaseous molecules.

Thus, it is likely (and is, indeed, the case) that lattice vibrations are not frozen out at room temperature, but, instead, make their full classical contribution to the molar specific heat of the solid.

If the lattice vibrations behave classically then, according to the equipartition theorem, each normal mode of oscillation has an associated mean energy in kT equilibrium at temperature T[(1/2)kT resides in the kinetic energy of the oscillation, and (1/2)kT resides in the potential energy]. Thus, the mean internal energy per mole of the solid is

$$\overline{E} = 3N\,k\,T = 3\nu RT.$$

It follows that the molar heat capacity at constant volume is

$$cv = \frac{1}{\nu}\left(\frac{\partial \overline{E}}{\partial T}\right)_v = 3R$$

for solids. This gives a value of 24.9 joules/mole/degree. In fact, at room temperature most solids (in particular, metals) have heat capacities which lie remarkably close to this value. This fact was discovered experimentally by Dulong and Petite at the beginning of the nineteenth century, and was used to make some of the first crude estimates of the molecular weights of solids (if we know the molar heat capacity of a substance then we can easily work out how much of it corresponds to one mole, and by weighing this amount, and then dividing the result by Avogadro's number, we can obtain an estimate of the molecular weight).

The experimental molar heat capacities c_p at constant pressure for various solids. The heat capacity at constant volume is somewhat less than the constant pressure value, but not by much, because solids are fairly incompressible.

It can be seen that *Dulong and Petite's law* (*i.e.*, that all solids have a molar heat capacities close to 24.9 joules/mole/degree) holds pretty well for metals. However, the law fails badly for diamond. This is not surprising. As is well-known, diamond is an extremely hard substance, so its intermolecular bonds must be very strong, suggesting that the force constant k is large. Diamond is also a fairly low density substance, so the mass involved in lattice vibrations is comparatively small.

Both these facts suggest that the typical lattice vibration frequency of diamond (w ~ $\sqrt{k/m}$) is high. In fact, the spacing between the different vibration energy levels (which scales like $\hbar w$) is sufficiently large in diamond for the vibrational degrees of freedom to be largely frozen out at room temperature.

Table 4: *Values of c_p (joules/mole/degree) for some solids at T = 298° K. From Reif.*

Solid	cp	Solid	c_p
Copper	24.5	Aluminium	24.4
Silver	25.5	Tin (white)	26.4
Lead	26.4	Sulphur (rhombic)	22.4
Zinc	25.4	Carbon (diamond)	6.1

Dulong and Petite's law is essentially a high temperature limit. The molar heat capacity cannot remain a constant as the temperature approaches absolute zero, since, by Equation, this would imply $S \to \infty$, which violates the third law of thermodynamics. We can make a crude model of the behaviour of c_v at low temperatures by assuming that all the normal modes oscillate at the same frequency, w, say. This approximation was first employed by Einstein in a paper published in 1907.

According to Equation, the solid acts like a set of $3N$ independent oscillators which, making use of Einstein's approximation, all vibrate at the same frequency. We can use the quantum mechanical result Equation for a single oscillator to write the mean energy of the solid in the form

$$\bar{E} = 3N\hbar w\left(\frac{1}{2} + \frac{1}{\exp(\beta\hbar w) - 1}\right).$$

The molar heat capacity is defined

$$cv = \frac{1}{v}\left(\frac{\partial \bar{E}}{\partial T}\right)_v = \frac{1}{v}\left(\frac{\partial \bar{E}}{\partial T}\right)_v \frac{\partial \beta}{\partial T} = -\frac{1}{vkT^2}\left(\frac{\partial \bar{E}}{\partial \beta}\right)_v,$$

giving

$$c_v = \frac{3N_A\hbar w}{kT^2} = \left[-\frac{\exp(\beta\hbar w)\hbar w}{[\exp(\beta\hbar w) - 1]^2}\right],$$

which reduces to

$$c_v = 3\,R = \left(\frac{\theta_E}{T}\right)^2 \frac{\exp(\theta_E / T)}{[\exp(\theta_E / T) - 1]^2}.$$

Here,

$$\theta_E = \frac{\hbar w}{k}$$

is called the *Einstein temperature*. If the temperature is sufficiently high that $T \gg \theta_E$ then $kT \gg \hbar\, w$, and the above expression reduces to $c_v = 3\,R$, after expansion of the exponential functions. Thus, the law of Dulong and Petite is recovered for temperatures significantly in excess of the Einstein temperature. On the other hand, if the temperature is sufficiently low that $T \ll \theta_E$ then the exponential factors in Equation become very much larger than unity, giving

$$c_v \sim 3\,R\frac{\theta_E}{T}\exp\,(-\theta_E / T).$$

So, in this simple model the specific heat approaches zero exponentially as $T \to 0$.

In reality, the specific heats of solids do not approach zero quite as quickly as suggested by Einstein's model when $T \rightarrow 0$. The experimentally observed low temperature behaviour is more like $c_v \propto T^3$. The reason for this discrepancy is the crude approximation that all normal modes have the same frequency. In fact, long wavelength modes have lower frequencies than short wavelength modes, so the former are much harder to freeze out than the latter (because the spacing between quantum energy levels, $\hbar w$, is smaller in the former case).

The molar heat capacity does not decrease with temperature as rapidly as suggested by Einstein's model because these long wavelength modes are able to make a significant contribution to the heat capacity even at very low temperatures.

A more realistic model of lattice vibrations was developed by the Dutch physicist Peter Debye in 1912. In the Debye model, the frequencies of the normal modes of vibration are estimated by treating the solid as an isotropic continuous medium. This approach is reasonable because the only modes which really matter at low temperatures are the long wavelength modes: *i.e.*, those whose wavelengths greatly exceed the interatomic spacing. It is plausible that these modes are not particularly sensitive to the discrete nature of the solid: *i.e.*, the fact that it is made up of atoms rather than being continuous.

Consider a sound wave propagating through an isotropic continuous medium. The disturbance varies with position vector r and time t like exp $[- i.r - wt]$, where the wave-vector k and the frequency of oscillation satisfy the dispersion relation for sound waves in an isotropic medium:

$$w = kc_s.$$

Here, c_s is the speed of sound in the medium. Suppose, for the sake of argument, that the medium is periodic in the x-, y-, and z-directions with periodicity lengths L_x, L_y, and L_z, respectively. In order to maintain periodicity we need

$$k_x (x + L_x) = k_x x + 2 \pi n_x,$$

where n_x is an integer. There are analogous constraints on k_y and k_z. It follows that in a periodic medium the components of the wave-vector are quantized, and can only take the values

$$k_x = \frac{2\pi}{L_x} n_x,$$

$$k_y = \frac{2\pi}{L_y} n_y,$$

$$k_z = \frac{2\pi}{L_z} n_z,$$

where n_x, n_y, n_z and are all integers. It is assumed that L_x, L_y, and L_z are macroscopic lengths, so the allowed values of the components of the wave-vector are very closely spaced. For given values of and k_y, the number of allowed values of k_x which lie in the range k_x to $k_x + dk_x$ is given by

$$\Delta n_x = \frac{L_x}{2\pi} dk_x,$$

It follows that the number of allowed values of k(*i.e.*, the number of allowed modes) when k_x lies in the range k_x to $k_x + dh_x$, k_y lies in the range k_y to $k_y + dk_y$, and k_z lies in the range k_z to $k_z + dh_z$, is

$$\rho d^3 k = \left(\frac{L_x}{2\pi} dk_x\right)\left(\frac{L_y}{2\pi} dk_y\right)\left(\frac{L_z}{2\pi} dk_z\right) = \frac{V}{(2\pi)^3} dk_x dk_y dk_z,$$

where $V = L_x L_y L_z$ is the periodicity volume, and $d^3k \equiv dk_x dk_y dk_z$. The quantity ρ is called the density of modes. Note that this density is independent of k, and proportional to the periodicity volume. Thus, the density of modes *per unit volume* is a constant independent of the magnitude or shape of the periodicity volume. The density of modes per unit volume when the *magnitude* of k lies in the range k to $k + dk$ is given by multiplying the density of modes per unit volume by the "volume" in k-space of the spherical shell lying between radii and $k + dk$. Thus,

$$\rho_k dk = \frac{4\pi k^2 dk}{(2\pi)^3} = \frac{k^2}{2\pi^2} dk.$$

Consider an isotropic continuous medium of volume V.

According to the above relation, the number of normal modes whose frequencies lie between w and $w + dw$ (which is equivalent to the number of modes whose k values lie in the range w/c_s to $w/c_s + d_w/c_s$) is

$$\sigma_c(w)dw = 3\frac{k^2V}{2\pi^2}dk = 3\frac{V}{2\pi^2 c_s{}^3}w^2 dw.$$

The factor of 3 comes from the three possible polarizations of sound waves in solids. For every allowed wavenumber (or frequency) there are two independent torsional modes, where the displacement is perpendicular to the direction of propagation, and one longitudinal mode, where the displacement is parallel to the direction of propagation.

Torsion waves are vaguely analogous to electromagnetic waves (these also have two independent polarizations). The longitudinal mode is very similar to the compressional sound wave in gases. Of course, torsion waves can not propagate in gases because gases have no resistance to deformation without change of volume.

The Debye approach consists in approximating the actual density of normal modes $\sigma(w)$ by the density in a continuous medium $\sigma_c(w)$, not only at low frequencies (long wavelengths) where these should be nearly the same, but also at higher frequencies where they may differ substantially. Suppose that we are dealing with a solid consisting of N atoms. We know that there are only $3N$ independent normal modes.

It follows that we must cut off the density of states above some critical frequency, w_D say, otherwise we will have too many modes. Thus, in the Debye approximation the density of normal modes takes the form

$$\sigma_D(\mathrm{w}) = \sigma_c(w) \quad \text{for } w < w_D$$
$$\sigma_D(\mathrm{w}) = 0 \text{ for } w < w_D.$$

Here, w_D is the *Debye frequency*. This critical frequency is chosen such that the total number of normal modes is $3N$, so

$$\int_0^\infty \sigma_D(w)dw = \int_0^{w_D} \sigma_c(w)dw = 3N.$$

Substituting Equation into the previous formula yields

$$\frac{3V}{2\pi^2}\int_0^{w_D} w^2 dw = \frac{V}{2\pi^2 c_s{}^3} w_D^3 = 3N.$$

This implies that

$$w_D = c_s\left(6\pi^2 \frac{N}{V}\right)^{1/3}.$$

Thus, the Debye frequency depends only on the sound velocity in the solid and the number of atoms per unit volume. The wavelength corresponding to the Debye frequency is $2\pi C_s/w_D$, which is clearly on the order of the interatomic spacing $a \sim (V/N)^{1/3}$. It follows that the cut-off of normal modes whose frequencies exceed the Debye frequency is equivalent to a cut-off of normal modes whose wavelengths are less than the interatomic spacing. Of course, it makes physical sense that such modes should be absent.

The actual density of normal modes in diamond with the density predicted by Debye theory. Not surprisingly, there is not a particularly strong resemblance between these two curves, since Debye theory is highly idealized. Nevertheless, both curves exhibit sharp cut-offs at high frequencies, and coincide at low frequencies. Furthermore, the areas under both curves are the same. Sufficient to allow Debye theory to correctly account for the temperature variation of the specific heat of solids at low temperatures.

We can use the quantum mechanical expression for the mean energy of a single oscillator, Equation, to calculate the mean energy of lattice vibrations in the Debye approximation. We obtain

$$\overline{E} = \int_0^\infty \sigma_D(w)\hbar w\left(\frac{1}{2}+\frac{1}{\exp(\beta\hbar w)-1}\right)dw.$$

According to Equation, the molar heat capacity takes the form

$$c_V = \frac{1}{vkT^2}\int_0^\infty \sigma_D(w)\hbar w\left[\frac{\exp(\beta\hbar w)\hbar w}{[\exp(\beta\hbar w)-1]^2}\right]dw.$$

Substituting in Equation, we find that

$$c_V = \frac{k}{v}\int_0^{w_D} \frac{\exp(\beta\hbar w)(\beta\hbar w)^2}{[\exp(\beta\hbar w)-1]^2}\frac{3V}{2\pi^2 C_s^3} w^2 dw,$$

giving

$$c_V = \frac{3V\,k}{2\pi^2 v(C_s\beta\hbar)^3}\int_0^{\beta\hbar w_D} \frac{\exp x}{(\exp x-1)^2} x^4 dx,$$

in terms of the dimensionless variable $x = \beta\hbar\, w$. According to Equation, the volume can be written

$$V = 6p^2 N\left(\frac{C_s}{w_D}\right)^3,$$

so the heat capacity reduces to

$$C_V = 3\,Rf_D(\beta\hbar w_D) = 3\,Rf_D(\theta_D/T)$$

where the *Debye function* is defined

$$f_D(y) \equiv \frac{3}{y^3}\int_0^y \frac{\exp x}{(\exp x-1)} x^4 dx.$$

We have also defined the *Debye temperature* θ_D as

$$k\theta_D = \hbar w_D.$$

Consider the asymptotic limit in which $T \gg \theta_D$. For small y, we can approximate $\exp x$ as $1 + x$ in the integrand of Equation, so that

$$f_D(y) \rightarrow \frac{3}{y^3}\int_0^y x^2 dx = 1.$$

Thus, if the temperature greatly exceeds the Debye temperature we recover the law of Dulong and Petite that $C_V = 3$ R. Consider, now, the asymptotic limit in which $T \ll \theta_D$. For large y,

$$\int_0^y \frac{\exp x}{(\exp x-1)^2} x^4 dx \simeq \int_0^\infty \frac{\exp x}{(\exp x-1)^2} x^4 dx = \frac{4\pi^4}{15}.$$

The latter integration is standard (if rather obscure), and can be looked up in any (large) reference book on integration. Thus, in the low temperature limit

$$f_D(y) \to \frac{4\pi^4}{5}\frac{1}{y^3}.$$

This yields

$$C_V \simeq \frac{12\pi^4}{5} R\left(\frac{T}{\theta_D}\right)^3$$

in the limit $T \ll \theta_D$: *i.e.*, C_V varies with temperature like T^3.

Table: Comparison of Debye temperatures (in degrees kelvin) obtained from the low temperature behaviour of the heat capacity with those calculated from the sound speed.

Solid	θ_D from low temp	θ_D from sound speed
NaCl	308	320
KCl	230	246
Ag	225	216
Zn	308	305

The fact that C_V goes like T^3 at low temperatures is quite well verified experimentally, although it is sometimes necessary to go to temperatures as low as 0.02 θ_D to obtain this asymptotic behaviour. Theoretically, θ_D should be calculable from Equation in terms of the sound speed in the solid and the molar volume. Debye temperatures evaluated by this means with temperatures obtained empirically by fitting the law Equation to the low temperature variation of the heat capacity.

It can be seen that there is fairly good agreement between the theoretical and empirical Debye temperatures. This suggests that the Debye theory affords a good, thought not perfect, representation of the behaviour of C_V in solids over the entire temperature range.

REFERENCES

- Statistical Physics, L. D. Landau and E. M. Lifshitz
- Statistical Mechanics, S-K Ma
- Quantum Mechanics, E. Merzbacher
- The Cluster Expansion, W. J. Mullin
- A Modern Course in Statistical Physics, L. E. Reichl.

6

Thermal Equilibrium and Zeroth Law

STATISTICAL DEFINITION OF THERMODYNAMIC VARIABLES

Our starting point is the idea that one can count the number of available states of a system. In principle, these are discrete quantum states. For a large system the states will be very closely spaced. The number of possible states with energy between E and $E + \delta E$ is

$$\Omega(E) = g(E)\, \delta E,$$

where $g(E)$ is the density of states. Next, consider a closed system with fixed volume V, number of particles N, and energy E. In order to avoid problems associated with the discreteness of the quantum states we take the energy to be specified within a tolerance δE. This tolerance should be chosen so that for a large system the precise value of δE does not matter. We do not know in which of the allowed states the system finds itself. In fact, our fundamental assumption is that at equilibrium our ignorance in this matter is complete, and that all the $\Omega(E, V, N)$possible states are equally likely, i.e. all memory of how the system was initially prepared is lost, except for the values of the energy, volume, and number of particles. We define the entropy as

$$S = k_B \ln \Omega(E, N, V).$$

Consider next an infinitesimally small change from an

equilibrium state E, V, N to another, slightly different, equilibrium state $E + dE$, $V + dV$, $N + dN$. The change in the entropy is then

$$dS = \frac{\partial S}{\partial E} dE + \frac{\partial S}{\partial V} dV + \frac{\partial S}{\partial n} dN.$$

The change in energy in this process is given by

$$dE = dQ + dU,$$

We distinguish between two forms of energy *heat* and *work*. Heat is a form of energy associated with random or *thermal* motion of atoms and molecules. Consider a gas of low density. The molecules will move in straight trajectories until they collide with other molecules or the walls of the gas container. After a few collisions it becomes practically impossible to relate the velocity and position of the molecules to the corresponding quantities at an earlier time.

The difficulty is not just the enormous amount of data required to describe a large number of particles. A more fundamental problem is the fact that after a few collisions the positions and the velocities of the particles become extremely sensitive to the initial conditions. A very similar situation occurs when throwing an unbiased die or tossing a coin. In principle, it should be possible to predict the outcome of the toss using Newton's laws and the initial velocity and position. In practice, the calculation will not be able to predict the behaviour of real coins, because initial conditions that give rise to radically different outcomes are so close together that the problem of specifying the intial conditions and parameters of the problem with sufficient accuracy becomes severe.

This type of motion has been described as *chaotic*. Each particle is just as likely to move in any direction as in any other, and the the speed of the particles is frequently changing. We also distinguish between the random motion of a molecule and bulk (ordered) movement.

An example of the latter is the flight of a solid object such as a pebble thrown in the air. We refer to changes in energy associated with bulk motion or transport of matter as *work*. In dQ is the heat *supplied to* the system and dU the work *done on* the system. The internal energy E is a state variable and its differential is exact, i.e. dE depends only on the initial and final state and is independent of the process leading to the change.

On the other hand "heat" and "work" are not state variables and the partition into heat and work depends on the process. Hence, the difference in notation: dE, but dQ and dU We have not yet defined the variables P, T and μ. We want to do this in such a way as to allow us to write

$$dE = TdS - PdV + \mu dN$$

or

$$dS = \frac{1}{T}dE + \frac{\mu}{T}dE + \frac{\mu}{T}dN + \frac{P}{T}dV.$$

We now define the temperature, pressure, and chemical potential as

$$T = \left(\frac{\partial S}{\partial E}\right)^{-1}_{N,V} \qquad \mu = -T\left(\frac{\partial S}{\partial N}\right)_{E,V} \qquad P = T\left(\frac{\partial S}{\partial V}\right)_{N,E}$$

It is important to note that our basic assumption is that all allowed states are equally likely. The second law of thermodynamics now becomes the statement that a closed system will tend to approach a macroscopic state which can be achieved the most possible ways. The conventional mathematical formulation of the second law on the other hand only becomes an essentially trivial matter of definition. We must next show that these definitions lead to familiar looking results- otherwise they would not be useful.

THE ZEROTH LAW OF THERMODYNAMICS

To establish the equivalence of our definitions and the conventional thermodynamic ones we shall make contact with the zeroth law of thermodynamics. This law has an analogy with mechanics, where in equilibrium the forces are balanced. In particular if two subsystems are in contact and in equilibrium:

$T_1 = T_2 \rightarrow$ thermal equilibrium

$P_1 = P_2 \rightarrow$ mechanical equilibrium

$\mu 1 = \mu_2\ T_2 \rightarrow$ chemical equilibrium

The zeroth law has a fairly straightforward statistical interpretation and this will allow us to begin to establish the equivalence between the statistical definitions and the conventional thermodynamic ones. Consider two systems that are free to exchange energy but are isolated from the rest of the universe by an ideal insulating surface. The particle numbers N_1, N_2 and volumes V_1, V_2 are fixed for each subsystem. The total energy will be constant under our assumptions and we assume further that the two subsystems are sufficiently weakly interacting that

$$E = E_1 + E_2,$$

where E_1 and E_2 are the energies of the subsystems. Assume that the densities of state $g(E)$, $g_1(E)$,$g_2(E)$ are coarse grained so that $\Omega = g(E)\delta E$ $\Omega_1 = g_1(E_1)\,\delta E$ $\Omega_2 = g_2(E_2)\,\delta E$, , . We then have

$$g(E) = \int dE_1 g_2(E - E_1) g_1(E_1),$$

If the subsystems are sufficiently large, the product $g_2(E-E_1)g_1(E_1)$ will be a sharply peaked function of E_1. From the definition of the entropy we note that it is a monotonically increasing function of g and that the product g_1g_2 will be at a maximum when the total entropy

$$S(E, E_1) = S_1(E_1) + S_2(E-E_1)$$

is at a maximum. The most likely value $\langle E_1\rangle$ of E_1 is the one for which

$$\frac{\partial S_1}{\partial E_1} + \frac{\partial S_2}{\partial E_2}\frac{\partial E_2}{\partial E_1} = 0.$$

Since $\partial E_2/\partial E_1$ we find using, that

$$\frac{1}{T_1} - \frac{1}{T_2} = 0,$$

or $T_1 = T_2$. The most probable partition of energy between the two systems is the one for which the two temperatures are the same. Consider next two subsystems that are separated by a movable wall. The two systems are free to exchange energy, but the number of particles is fixed in each subsystem and the total volume $V = V_1 + V_2$ is constant. We write $E = E_1 + E_2$. The density of allowed states for the total system is then

$$g(E,V) = \sum_{\text{volume settings}} \int dE_1 g_1(E_1,V_1) g_{21}(E-E_1, V-V_1)$$

The integrand is sharply peaked for large systems and takes on its maximum value when $S_1(E_1,V_1)+S_2(E-E_1,V-V_1)=max$. Differentiation using

$$\left(\frac{\partial S}{\partial E}\right)_{VN} = \frac{1}{T}; \left(\frac{\partial S}{\partial V}\right)_{EN} = \frac{P}{T}$$

implies that it is overwhelmingly probable that the system will be near a state for which

$$\frac{1}{T_1} = \frac{1}{T_2}; \frac{P_1}{T_1} = \frac{P_2}{T_2}$$

or $T_1 = T_2, P_1 = P_2$. Conventionally, one would say that the pressure in the two compartments must be equal at equilibrium because the forces have to be in balance. The argument now being made is quite different, there are no forces, instead the movable wall is guided to its equilibrium by the invisible hand of the law of large numbers. Similarly, consider two systems 1 and 2 which are free to exchange particles and energy. It is easy to show that the most probable configuration is the one for which $T_1 = T_2$, $\mu_1 = \mu_2$.

BOLTZMANN FACTOR

Consider now a system in contact with a heat bath, or reservoir. System 1 is the one we are interested in, and we want to find the probability $P(E_1)$ that it has energy E_1. We assume that system 2 is much larger than 1, so that $E_1 \ll E = E_1 + E_2$. Another way of putting this is to say that the heat capacity C_2 of system 2 is very large. We assume that all compatible microstates are equally likely. We have

$$P(E_1)dE_1 = \frac{g_1(E_1)g_2(E-E_1)dE_1}{\int dE_1 g_1(E_1)g_2(E-E_1)}$$

From the definition of entropy

$$g_2 = \frac{1}{\delta E}\exp\left[\frac{S_2(E-E_1}{k_B}\right]$$

We expand in a Taylor series

$$S_2(E - E_1) = S_2(E) - E_1 \frac{\partial S_2}{\partial E} + \frac{E_1^2}{2} \frac{\partial^2 S_2}{\partial E^2} + ...$$

With T the temperature of the heat bath and C its heat capacity, the partial derivatives are given by

$$\frac{\partial S_2}{\partial E} = \frac{1}{T}$$

$$\frac{\partial^2 S_2}{\partial E^2} = \frac{\partial \frac{1}{T}}{\partial E} = -\frac{1}{T^2}\frac{\partial T}{\partial E} = -\frac{1}{T^2 C}$$

Since $E_1 << TC$ we neglect the last term in giving

$$g_2 = const.\exp\left[\frac{-E_1}{k_B T}\right] = const.e^{-\beta E_1}$$

where we define $\beta = 1/(k_B T)$. We conclude

The factor $e^{-\beta E1}$ is the *Boltzmann factor*. When a system is in contact with a heat bath at a certain temperature, all possible microstates of the system are no longer equally likely. Instead, the Boltzmann factor acts as a weight factor biasing the distribution towards states with lower energy.

PARTITION FUNCTION AND THE CANONICALDISTRIBUTION

The constant in eqnaution can be determined by *normalizing* the probability distribution i.e. requiring that

$$\int p(E)dE = 1$$

Let us define the *canonical partition function* (α = microstate)

$$Z_c = \sum_{\alpha} e^{-\beta E(\alpha)} = \int dE g(E) e^{-\beta E}$$

We find that the probability $p(\alpha)$ that a state is in a given microstate

$$p(\alpha) = \frac{1}{Z_c} e^{-\beta E}$$

If $x(\alpha)$ is the value of some physical property in microstate α, and $E(\alpha)$ the energy of this state then the *canonical ensemble average* is given by

$$\langle x\rangle = \frac{1}{Z_c}\sum_{\alpha} x(\alpha)e^{-\beta E(\alpha)}$$

Equation is a very useful formula, and we will give many examples of its use.

HELMHOLTZ FREE ENERGY

For an isolated system $S = S(E,V,N)$, with E,V,N *independent variables*. For a system in contact with a heat bath at a given temperature, T becomes an independent variable, or control parameter. The energy E and entropy S will then fluctuate about their mean values and . E and S become *dependent variables* given by equations of state. The change of variables is handled most efficiently by introducing he *Helmholtz free energy*. In thermodynamics it is defined as

$$F = E - TS$$

Imagine a reversible process which takes the system from one equilibrium state to another

$$dE = TdS - PdV + \mu dN = dE\ (S, V, N)$$

$$dF = dE - TdS - SdT = -PdV + \mu dN = dF\ (T, V, N)$$

We see that the Helmholtz free energy is should be considered to be dependent on the control variables T, V, N. We have

$$S = -\frac{\partial F}{\partial T}$$

$$P = -\frac{\partial F}{\partial T}$$

$$\mu = -\frac{\partial F}{\partial N}$$

In statistical mechanics we define the Helmholtz free energy as

$$\mathrm{A} = \langle \mathrm{E}\rangle - T\langle \mathrm{S}\rangle = \langle \mathrm{F}\rangle$$

We wish to show that for a large system

$$Z_c \int eEg(E)e^{-\beta E}$$

Proof: The canonical partition function is

$$= \int \frac{dE}{\delta E} \exp\{\beta[E - TS(E, V, N)]\}$$

We evaluate this integral using the *saddle point method.* Almost all the contribution to the integral will come from values of E near $E = \langle E \rangle$ the value for which E-$T(S, E, V, N)$ = minimum. We let $S(\langle E \rangle, V, N) = \langle S \rangle$ and

$$E - TS \simeq \langle E \rangle - T\langle S \rangle - \frac{1}{2} T(E - \langle E \rangle)^2 \frac{\partial^2 S}{\partial E^2} + ...$$

Substituting Equation we obtain

$$Z_c \simeq \exp[-\beta(\langle E \rangle - T\langle S \rangle) - T\langle S \rangle)] \int \frac{dE}{\delta E} \exp\left\{ \frac{-(E - \langle E \rangle)^2}{2Ck_B T^2} \right\}$$

Using we find

$$Z_c \simeq \frac{\sqrt{2\pi k_B T^2 C}}{\delta E} \exp[-\beta(\langle E \rangle - T\langle S \rangle)]$$

and from

$$A = -k_B T \ln Z_c = \langle E \rangle - T\langle S \rangle - k_B T \ln \left[\frac{\sqrt{2\pi k_B T^2 C}}{\delta E} \right]$$

The last term in will be small compared to the first two terms for a large system, and it is possible to choose the tolerance δE so that it is identically zero.

BOLTZMANN DISTRIBUTIONS

We have gained some understanding of the macroscopic properties of the air around us. For instance, we know something about its internal energy and specific heat capacity. How can we obtain some information about the statistical properties of the molecules which make up air? Consider a specific molecule: it constantly collides with its immediate neighbour molecules, and occasionally bounces off the walls of the room.

These interactions "inform" it about the macroscopic state of the air, such as its temperature, pressure, and volume. The statistical distribution of the molecule over its own particular microstates must be consistent with this macrostate. In other words, if we have a large group of such molecules with similar statistical distributions, then they must be equivalent to air with the appropriate macroscopic properties. So, it ought to be possible to calculate the probability distribution of the molecule over its microstates from a knowledge of these macroscopic properties.

We can think of the interaction of a molecule with the air in a classroom as analogous to the interaction of a small system A in thermal contact with a heat reservoir A'. The air acts like a heat reservoir because its energy fluctuations due to any interactions with the molecule are far too small to affect any of its macroscopic parameters. Let us determine the probability P_r of finding system A in one particular microstate r of energy Er when it is thermal equilibrium with the heat reservoir A'.

As usual, we assume fairly weak interaction between A and A', so that the energies of these two systems are additive. The energy of A is not known at this stage. In fact, only the *total* energy of the combined system $A^{(0)} = A + A'$ is known. Suppose that the total energy lies in the range $E^{(0)}$ to $E^{(0)} + \delta E$. The overall energy $A^{(0)}$ is constant in time, since is assumed to be an isolated system, so

$$E_r + E' = E^{(0)},$$

where E' denotes the energy of the reservoir A'. Let $\Omega'(E')$ be the number of microstates accessible to the reservoir when its energy lies in the range E' to $E' + \delta E$. Clearly, if system A has an energy E_r then the reservoir A' must have an energy close to $E = E^{(0)} - E_r$. Hence, since A is in *one* definite state (*i.e.*, state r), and the total number of states accessible to A' is $\Omega'(E^{(0)} - E_r)$, it follows that the total number of states accessible to the combined system is simply $\Omega'(E^{(0)} - E_r)$. The principle of equal *a priori* probabilities tells us the the probability of occurrence of a particular situation is proportional to the number of accessible microstates. Thus,

$$Pr = C' \, \Omega'(E^{(0)} - E_r)$$

where is a constant of proportionality which is independent of r. This constant can be determined by the normalization condition

$$\sum_r P_r = 1,$$

where the sum is over all possible states of system A, irrespective of their energy.

Let us now make use of the fact that system A is far smaller than system A'. It follows that $E_r \ll E^{(0)}$, so the slowly varying logarithm of P_r can be Taylor expanded about $E' = E^{(0)}$. Thus,

$$\ln P_r = \ln C' + \ln \Omega'(E^{(0)}) - \left[\frac{\partial \ln \Omega'}{\partial E'}\right]_0 Er + \cdots.$$

Note that we must expand ln P, rather than P_r itself, because the latter function varies so rapidly with energy that the radius of convergence of its Taylor series is far too small for the series to be of any practical use. The higher order terms in Equation can be safely neglected, because $E_r \ll E^{(0)}$. Now the derivative

$$\left[\frac{\partial \ln \Omega'}{\partial E'}\right]_0 \equiv \beta.$$

is evaluated at the fixed energy $E' = E^{(0)}$ and is, thus, a constant independent of the energy E_r of A. In fact, we know, that this derivative is just the temperature parameter $\beta = (kT)^{-1}$ characterizing the heat reservoir A'. Hence, Equation becomes

$$\ln P_r = \ln C' + \ln \Omega'(E^{(0)}) - \beta E_r,$$

giving

$$P_r = C \exp(-\beta E_r),$$

where C is a constant independent of r. The parameter C is determined by the normalization condition, which gives

$$C^{-1} = \sum_r \exp(-\beta E_r)$$

so that the distribution becomes

$$P_r = \frac{\exp(-\beta E_r)}{\sum_r \exp(-\beta E_r)}.$$

This is known as the *Boltzmann probability distribution*, and is undoubtably the most famous result in statistical physics.

The Boltzmann distribution often causes confusion. People who are used to the principle of equal *a priori* probabilities, which says that all microstates are equally probable, are understandably surprised when they come across the Boltzmann distribution which says that high energy microstates are markedly less probable then low energy states.

However, there is no need for any confusion. The principle of equal *a priori* probabilities applies to the *whole* system, whereas the Boltzmann distribution only applies to a small part of the system. The two results are perfectly consistent. If the small system is in a microstate with a comparatively high energy E_r then the rest of the system (*i.e.*, the reservoir) has a slightly lower energy E' than usual (since the overall energy is fixed).

The number of accessible microstates of the reservoir is a very strongly increasing function of its energy. It follows that when the small system has a high energy then significantly less states than usual are accessible to the reservoir, and so the number of microstates accessible to the overall system is reduced, and, hence, the configuration is comparatively unlikely.

The strong increase in the number of accessible microstates of the reservoir with increasing E' gives rise to the strong (*i.e.*, exponential) decrease in the likelihood of a state of the small system with increasing E_r. The exponential factor $\exp(-\beta E_r)$ is called the *Boltzmann factor.*

The Boltzmann distribution gives the probability of finding the small system A in *one* particular state of energy E_r. The probability $P(E)$ that A has an energy in the small range between E and $E + \delta E$ is just the sum of all the probabilities of the states which lie in this range. However, since each of these states has approximately the same Boltzmann factor this sum can be written

$$P(E) = C\,\Omega(E)\exp(-\beta E),$$

where $\Omega(E)$ is the number of microstates of A whose energies lie in the appropriate range. Suppose that system A is itself a large system, but still very much smaller than system A'. For a large system, we expect $\Omega(E)$ to be a very rapidly increasing function of energy, so the probability $P(E)$ is the product of a rapidly increasing function of E and another rapidly decreasing function (*i.e.*, the Boltzmann factor).

This gives a sharp maximum of $P(E)$ at some particular value of the energy. The larger system A, the sharper this maximum becomes. Eventually, the maximum becomes so sharp that the energy of system A is almost bound to lie at the most probable energy. As usual, the most probable energy is evaluated by looking for the maximum of $\ln P$, so

$$\frac{\partial \ln P}{\partial E} = \frac{\partial \ln \Omega}{\partial E} - \beta = 0,$$

giving

$$\frac{\partial \ln \Omega}{\partial E} - \beta$$

Of course, this corresponds to the situation in which the temperature of A is the same as that of the reservoir. This is a result which we have seen before. Note, however, that the Boltzmann distribution is applicable no matter how *small* system A is, so it is a far more general result than any we have previously obtained.

PARAMAGNETISM

The simplest microscopic system which we can analyze using the Boltzmann distribution is one which has only two possible states (there would clearly be little point in analyzing a system with only one possible state). Most elements, and some compounds, are *paramagnetic*: *i.e.*, their constituent atoms, or molecules, possess a permanent magnetic moment due to the presence of one or more unpaired electrons. Consider a substance whose constituent particles contain only one unpaired electron.

Such particles have spin 1/2, and consequently possess an intrinsic magnetic moment μ. According to quantum mechanics, the magnetic moment of a spin 1/2 particle can point either parallel or antiparallel to an external magnetic field B. Let us determine the mean magnetic moment $\bar{\mu}_B$ (in the direction of B) of the constituent particles of the substance when its absolute temperature is T. We assume, for the sake of simplicity, that each atom (or molecule) only interacts weakly with its neighbouring atoms. This enables us to focus attention on a single atom, and treat the remaining atoms as a heat bath at temperature .

Our atom can be in one of two possible states: the (+) state in which its spin points up (*i.e.*, parallel to B), and the (–) state in which its spin points down (*i.e.*, antiparallel to B). In the (+) state, the atomic magnetic moment is parallel to the magnetic field, so that $\mu_B = \mu$. The magnetic energy of the atom is $\in_+ = -\mu_B$. In the (–) state, the atomic magnetic moment is antiparallel to the magnetic field, so that $\mu_B = -\mu$. The magnetic energy of the atom is $\in\_ = \mu_B$.

According to the Boltzmann distribution, the probability of finding the atom in the (+) state is

$$P_+ = C\exp(-\beta\in_+) = C\exp(\beta\,\mu\,B),$$

where C is a constant, and $\beta = (kT)^{-1}$. Likewise, the probability of finding the atom in the (–) state is

$$P\_ = C\exp(-\beta\in\_) = C\exp(\beta\,\mu\,B).$$

Clearly, the most probable state is the state with the lowest energy [*i.e.*, the (+) state]. Thus, the mean magnetic moment points in the direction of the magnetic field (*i.e.*, the atom is more likely to point parallel to the field than antiparallel).

It is clear that the critical parameter in a paramagnetic system is

$$y \equiv \beta\,\mu\,B = \frac{\mu B}{kT}.$$

This parameter measures the ratio of the typical magnetic energy of the atom to its typical thermal energy. If the thermal energy greatly exceeds the magnetic energy then $y \ll 1$, and the probability that the atomic moment points parallel to the magnetic field is about the same as the probability that it points antiparallel. In this situation, we expect the mean atomic moment to be small, so that $\bar{\mu}_B \simeq 0$.

On the other hand, if the magnetic energy greatly exceeds the thermal energy then $y \gg 1$, and the atomic moment is far more likely to point parallel to the magnetic field than antiparallel. In this situation, we expect $\bar{\mu}_B \simeq \mu$.

Let us calculate the mean atomic moment $\bar{\mu}_B$. The usual definition of a mean value gives

$$\bar{\mu}_B = \frac{P_+ + \mu + P\_(-\mu)}{P_+ + P\_} = \mu\frac{\exp(\beta\,\mu\,B) - \exp(-\beta\mu\,B)}{\exp(\beta\,\mu\,B) + \exp(-\beta\mu\,B)}.$$

This can also be written

$$\bar{\mu}_B = \mu \tan h \frac{\mu B}{kT},$$

where the hyperbolic tangent is defined

$$\tan h\ y \equiv \frac{\exp(y) - \exp(-y)}{\exp(y) + \exp(-y)}.$$

For small arguments, $y \ll 1$,

$$\tan h\ y \simeq \frac{y^3}{3} + \cdots,$$

whereas for large arguments, $y \gg 1$,

$$\tan h\ y \simeq 1.$$

It follows that at comparatively high temperatures, $kT \gg \mu_B$,

$$\bar{\mu}_B \simeq \frac{\mu^2 B}{kT},$$

whereas at comparatively low temperatures, $kT \ll \mu B$,

$$\bar{\mu}_B \simeq \mu.$$

Suppose that the substance contains N_0 atoms (or molecules) per unit volume. The *magnetization* is defined as the mean magnetic moment per unit volume, and is given by

$$\bar{M}_0 = N_0 \bar{\mu}_B.$$

At high temperatures, $kT \gg \mu$B, the mean magnetic moment, and, hence, the magnetization, is proportional to the applied magnetic field, so we can write

$$\bar{M}_0 \simeq \chi B,$$

where χ is a constant of proportionality known as the *magnetic susceptibility*. It is clear that the magnetic susceptibility of a spin 1/2 paramagnetic substance takes the form

$$\chi = \frac{N_0 \mu^2}{kT}.$$

The fact that $\chi \propto T^{-1}$ is known as *Curie's law*, because it was discovered experimentally by Pierre Curie at the end of the nineteenth century. At low temperatures, $kT \ll \mu B$,

$$\bar{M}_0 \to N_0 \mu,$$

so the magnetization becomes independent of the applied field. This corresponds to the maximum possible magnetization, where all atomic moments are lined up parallel to the field. The breakdown of the $\bar{M}_0 \propto B$ law at low temperatures (or high magnetic fields) is known as *saturation*.

The above analysis is only valid for paramagnetic substances made up of spin one-half (J = 1/2) atoms or molecules. However, the analysis can easily be generalized to take account of substances whose constituent particles possess higher spin (*i.e.*, J > 1/2). The experimental and theoretical magnetization versus field-strength curves for three different substances made up of spin 3/2, spin 5/2, and spin 7/2 particles, showing the excellent agreement between the two sets of curves. Note that, in all cases, the magnetization is proportional to the magnetic field-strength at small field-strengths, but saturates at some constant value as the field-strength increases.

The previous analysis completely neglects any interaction between the spins of neighbouring atoms or molecules. It turns out that this is a fairly good approximation for paramagnetic substances. However, for *ferromagnetic* substances, in which the spins of neighbouring atoms interact very strongly, this approximation breaks down completely. Thus, the above analysis *does not* apply to ferromagnetic substances.

MEAN VALUES

Consider a system in contact with a heat reservoir. The systems in the representative ensemble are distributed over their accessible states in accordance with the Boltzmann distribution. Thus, the probability of occurrence of some state with energy E_r is given by

$$P_r = \frac{\exp(-\beta E_r)}{\sum_r \exp(-\beta E_r)}.$$

The mean energy is written

$$\overline{E} = \frac{\sum_r \exp(-\beta\, E_r) E_r}{\sum_r \exp(-\beta\, E_r)},$$

where the sum is taken over all states of the system, irrespective of their energy. Note that

$$\sum_r \exp(-\beta\, E_r) E_r = -\sum_r \frac{\partial}{\partial\beta} \exp(-\beta E_r) = -\frac{\partial z}{\partial\beta},$$

where

$$Z = \sum_r \exp(-\beta\, E_r)$$

It follows that p

$$\overline{E} = -\frac{1}{Z}\frac{\partial Z}{\partial\beta} = -\frac{\partial \ln Z}{\partial\beta}.$$

The quantity Z, which is defined as the sum of the Boltzmann factor over all states, irrespective of their energy, is called the *partition function*. We have just demonstrated that it is fairly easy to work out the mean energy of a system using its partition function. In fact, as we shall discover, it is easy to calculate virtually any piece of statistical information using the partition function.

Let us evaluate the variance of the energy. We know that

$$\overline{(\Delta E)^2} = \overline{E^2} - \overline{E}^2$$

Now, according to the Boltzmann distribution,

$$\overline{E^2} = \frac{\sum_r \exp(-\beta E_r) E_r^{\;2}}{\sum_r \exp(-\beta E_r)}.$$

However,

$$\sum_r \exp(-\beta E_r) E_r^{\;2} = -\frac{\partial}{\partial\beta}\left[\sum_r \exp(-\beta E_r) E_r\right] = \left(\frac{\partial}{\partial\beta}\right)^2 \left[\sum_r \exp(-\beta E_r)\right].$$

Hence,

$$\overline{E^2} = \frac{1}{Z}\frac{\partial^2 Z}{\partial \beta^2}.$$

We can also write

$$\overline{E^2} = \frac{\partial}{\partial \beta}\left(\frac{1}{Z}\frac{\partial Z}{\partial \beta}\right) + \frac{1}{Z^2}\left(\frac{\partial Z}{\partial \beta}\right)^2 = -\frac{\partial \overline{E}}{\partial \beta} + \overline{E}^2,$$

where use has been made of Equation

$$\overline{(\Delta E)^2} = -\frac{\partial \overline{E}}{\partial \beta} = \frac{\partial^2 \ln Z}{\partial \beta^2}.$$

Thus, the variance of the energy can be worked out from the partition function almost as easily as the mean energy. Since, by definition, a variance can never be negative, it follows that $\partial \overline{E} / \partial \beta \leq 0$, or, equivalently, $\partial \overline{E} / \partial T \geq 0$. Hence, the mean energy of a system governed by the Boltzmann distribution always increases with temperature. Suppose that the system is characterized by a single external parameter x (such as its volume). The generalization to the case where there are several external parameters is obvious. Consider a quasi-static change of the external parameter from x to $x + dx$. In this process, the energy of the system in state changes by

$$\delta E_r = \frac{\partial E}{\partial x} dx.$$

The macroscopic work dW done by the system due to this parameter change is

$$đW = \frac{\sum_r \exp(-\beta\, E_r)(-\partial E_r / \partial x\, dx)}{\sum_r \exp(-\beta\, E_r)}.$$

In other words, the work done is minus the average change in internal energy of the system, where the average is calculated using the Boltzmann distribution. We can write

$$\sum_r \exp(-\beta E_r)\frac{\partial E_r}{\partial x} = -\frac{1}{\beta}\frac{\partial}{\partial x}\left[\sum_r \exp(-\beta E_r)\right] = -\frac{1}{\beta}\frac{\partial Z}{\partial x},$$

which gives

$$đW = -\frac{1}{\beta Z}\frac{\partial Z}{\partial x}dx = \frac{1}{\beta}\frac{\partial \ln Z}{\partial x}dx.$$

We also have the following general expression for the work done by the system

$$đW = \bar{X}\, dx,$$

where

$$\bar{X} = -\overline{\frac{\partial E_r}{\partial x}},$$

is the mean generalized force conjugate to x. It follows that

$$\bar{X} = \frac{1}{\beta}\frac{\partial \ln Z}{\partial x}.$$

Suppose that the external parameter is the volume, so $x = V$. It follows that

$$đW = \bar{p}dV = \frac{1}{\beta}\frac{\partial \ln Z}{\partial V}dV$$

and

$$\bar{p} = \frac{1}{\beta}\frac{\partial \ln Z}{\partial V}.$$

Since the partition function is a function of β and V (the energies E_r depend on V), it is clear that the above equation relates the mean pressure $\bar{p}$ to T (via $\beta = 1/kT$) and V.

In other words, the above expression is the *equation of state*. Hence, we can work out the pressure, and even the equation of state, using the partition function.

PARTITION FUNCTIONS

It is clear that all important macroscopic quantities associated with a system can be expressed in terms of its partition function Z. Let us investigate how the partition function is related to thermodynamical quantities. Recall that Z is a function of both β and x (where x is the single external parameter). Hence, $Z = Z(\beta, x)$, and we can write

$$d\ln Z = \frac{\partial \ln Z}{\partial x}dx + \frac{\partial \ln Z}{\partial \beta}d\beta.$$

Consider a quasi-static change by which x and b change so slowly that the system stays close to equilibrium, and, thus, remains distributed according to the Boltzmann distribution. If follows from Equations that

$$d\ln Z = \beta đW - \bar{E}d\beta.$$

The last term can be rewritten

$$d\ln Z = \beta đW - d(\bar{E}\beta) + \beta d\bar{E},$$

giving

$$d(\ln Z + \beta\bar{E}) = \beta(đW + d\bar{E}) \equiv \beta đQ.$$

The above equation shows that although the heat absorbed by the system $đQ$ is not an exact differential, it becomes one when multiplied by the temperature parameter β. This is essentially the second law of thermodynamics. In fact, we know that

$$dS = \frac{đQ}{T}.$$

Hence,

$$S \equiv k(\ln Z + \beta\bar{E}).$$

This expression enables us to calculate the entropy of a system from its partition function. Suppose that we are dealing with a system $A^{(0)}$consisting of two systems A and A'which only interact weakly with one another. Let each state of A be denoted by an index and have a corresponding energy E_r. Likewise, let each state of A' be denoted by an index and have a corresponding energy E_s'. A state of the combined system $A^{(0)}$ is then denoted by two indices r and s. Since A and A' only interact weakly their energies are additive, and the energy of state is

$$E_{rs}^{(0)} = E_r + E_x'.$$

By definition, the partition function of $A^{(0)}$ takes the form

$$Z^{(0)} = \sum_{r,s} \exp[-\beta E_{rs}^{(0)}]$$

$$= \sum_{r,s} \exp(-\beta[E_r + E_s'])$$

$$= \sum_{r,s} \exp(-\beta E_r)\exp(-\beta E_s'])$$

$$= \left[\sum_r \exp(-\beta E_r)\right]\left[\sum_s \exp(-\beta E_s')\right].$$

Hence,

$$Z^{(0)} = Z\,Z',$$

giving

$$\ln Z^{(0)} = \ln Z + \ln Z',$$

where Z and Z' are the partition functions of A and A', respectively. It follows from Equations that the mean energies of $A^{(0)}$, A, and A' are related by

$$\overline{E}^{(0)} = \overline{E} + \overline{E}'.$$

It also follows from Equations that the respective entropies of these systems are related via

$$S^{(0)} = S + S'.$$

Hence, the partition function tells us that the extensive thermodynamic functions of two weakly interacting systems are simply additive. It is clear that we can perform statistical thermodynamical calculations using the partition function Z instead of the more direct approach in which we use the density of states Ω. The former approach is advantageous because the partition function is an *unrestricted* sum of Boltzmann factors over all accessible states, irrespective of their energy, whereas the density of states is a *restricted* sum over all states whose energies lie in some narrow range. In general, it is far easier to perform an unrestricted sum than a restricted sum. Thus, it is generally easier to derive statistical thermodynamical results using Z rather than Ω, although Ω has a far more direct physical significance than Z.

REFERENCES

- Heat and Thermodynamics, M. W. Zemansky
- Modern Thermodynamics, D. Kondepundi and I. Prigogine
- Noise and Fluctuations, D. K. C. MacDonald
- Introduction to Metallurgical Thermodynamics by David R. Gaskell.
- Introductory Statistical Mechanics, R. Bowley and M. Sánchez.

7

The First Law of Thermodynamics

Thermodynamics is a science and, more importantly, an engineering tool used to describe processes that involve changes in temperature, transformation of energy, and the relationships between heat and work.

It can be regarded as a generalization of an enormous body of empirical evidence. It is extremely general: there are no hypotheses made concerning the structure and type of matter that we deal with.

It is used to describe the performance of propulsion systems, power generation systems, and refrigerators, and to describe fluid flow, combustion, and many other phenomena. The focus of thermodynamics in aerospace engineering is on the production of work, often in the form of kinetic energy or shaft power, from different sources of heat.

For the most part the heat will be the result of combustion processes, but this is not always the case. The course content can be viewed in terms of a "propulsion chain", where a progression from an energy source to useful propulsive work (thrust power of a jet engine).

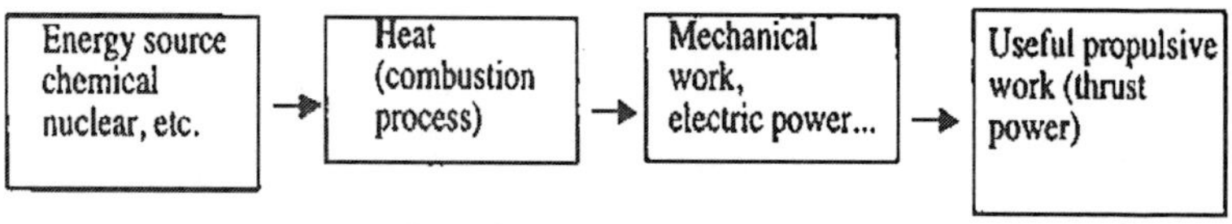

Fig. The Propulsion Chain

FUNDAMENTAL IDEAS OF THERMODYNAMICS

As with all sciences, thermodynamics is concerned with the mathematical modeling of the real world. In order that the mathematical deductions are consistent, we need some precise definitions of the basic concepts. Several of these will be further amplified in the lectures and in other handouts. If you need additional information or examples concerning these topics, they are described clearly and in-depth. Thermodynamics plays an important role in our understanding of electrochemical processes. It can tell us whether a given redox reaction is spontaneous and therefore whether it is able to provide useful electrical energy.

Thermodynamics also describes how to add reduction potentials to determine the cell potential for a galvanic or electrolytic cell. It also permits us to add reduction potentials of two reductions to calculate the potential of a new, or even theoretical, reduction half-reaction. Thermodynamics provides further insight into electrochemical cells at non-standard conditions in its derivation of the Nernst Equation. The Nernst Equation allows us to calculate the cell potential at any conditions and suggests the construction of concentration cells such as pH meters or other ion-selective electrodes.

Combining ideas from thermodynamics, stoichiometry, and basic electrical theory makes this section the most important to understand if you wish to become proficient at doing electrochemistry problems. Such problems tend to be among the more difficult ones on any exam because they cut across many fields in chemistry and require deep analytical thought and an thorough understanding of electrochemical cells. In this section, you will get to know and love such formulae as $\Delta G° = -nFE°$ (useful for converting free energy and potentials) and $E = E° - (RT/nF) \ln Q$ (the Nernst Equation).

THE CONTINUUM MODEL

Matter may be described at a molecular (or microscopic) level using the techniques of statistical mechanics and kinetic theory. For engineering purposes, however, we want "averaged" information, i.e., a macroscopic, not a microscopic, description. There are two reasons for this. First, a microscopic description of an engineering device may produce too much information to manage. 1 mm^3 of air at standard temperature and pressure contains 10^{16} molecules, each of which has a position and a velocity.

Typical engineering applications involve more than 10^{20} molecules. Second, and more importantly, microscopic positions and velocities are generally not useful for determining how macroscopic systems will act or react unless, for instance, their total effect is integrated.

We therefore neglect the fact that real substances are composed of discrete molecules and model matter from the start as a smoothed-out continuum. The information we have about a continuum represents the microscopic information averaged over a volume. Classical thermodynamics is concerned only with continua.

THE CONCEPT OF A "SYSTEM"

A thermodynamic system is a quantity of matter of fixed identity, around which we can draw a boundary. The boundaries may be fixed or moveable. Work or heat can be transferred across the system boundary. Everything outside the boundary is the surroundings.

When working with devices such as engines it is often useful to define the system to be an identifiable volume with flow in and out. This is termed a control volume.

A closed system is a special class of system with boundaries that matter cannot cross. Hence the principle of the conservation of mass is automatically satisfied whenever we employ a closed system analysis. This type of system is sometimes termed a control mass.

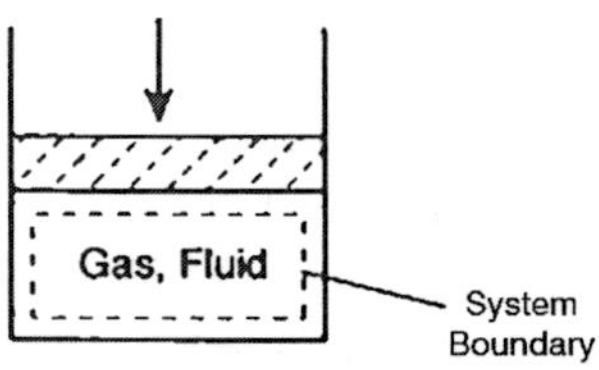

Fig. Piston (Boundary) and Gas (System)

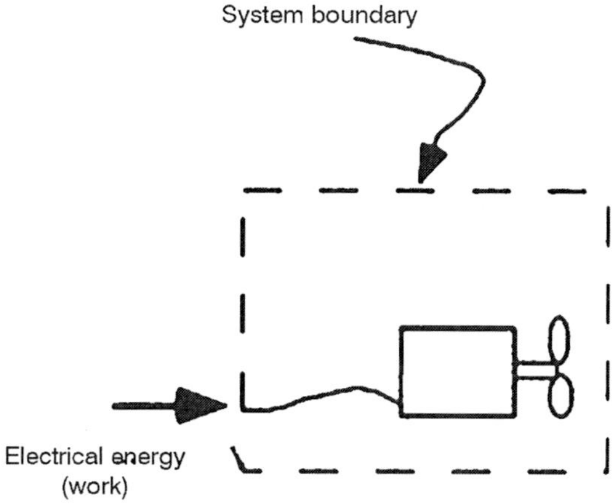

Fig. Boundary Around Electric Motor (System)

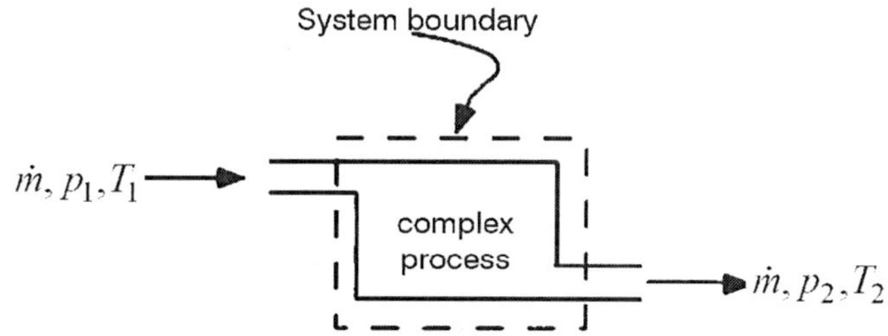

Fig. Sample Control Volume

THE CONCEPT OF A "STATE"

The thermodynamic state of a system is defined by specifying values of a set of measurable properties sufficient to determine all other properties. For fluid systems, typical properties are pressure, volume and temperature. More complex systems may require the specification of more unusual properties. As an example, the state of an electric battery requires the specification of the amount of electric charge it contains.

Properties may be extensive or intensive. Extensive properties are additive. Thus, if the system is divided into a number of sub-systems, the value of the property for the whole system is equal to the sum of the values for the parts. Volume is an extensive property. Intensive properties do not depend on the quantity of matter present. Temperature and pressure are intensive properties.

Specific properties are extensive properties per unit mass and are denoted by lower case letters.

$$\text{Specific volume} = V/m = v.$$

Specific properties are intensive because they do not depend on the mass of the system.

The properties of a simple system are uniform throughout. In general, however, the properties of a system can vary from point to point. We can usually analyze a general system by sub-dividing it (either conceptually or in practice) into a number of simple systems in each of which the properties are assumed to be uniform.

It is important to note that properties describe states only when the system is in equilibrium.

THE CONCEPT OF "EQUILIBRIUM"

The state of a system in which properties have definite, unchanged values as long as external conditions are unchanged is called an equilibrium state.

[Mechanical Equilibrium]

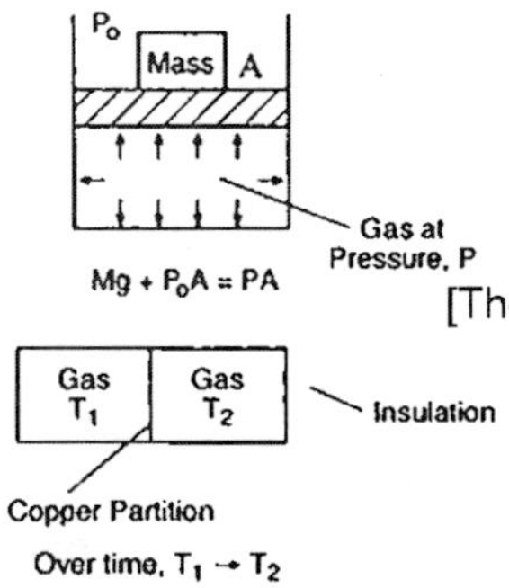

[Thermal Equilibrium]

Fig Equilibrium

A system in thermodynamic equilibrium satisfies:

- Mechanical equilibrium (no unbalanced forces);
- Thermal equilibrium (no temperature differences);
- Chemical equilibrium.

THE CONCEPT OF A "PROCESS"

If the state of a system changes, then it is undergoing a process. The succession of states through which the system passes defines the path of the process. If, at the end of the process, the properties have returned to their original values, the system has undergone a cyclic process or a cycle. Note that even if a system has returned to its original state and completed a cycle, the state of the surroundings may have changed.

QUASI-EQUILIBRIUM PROCESSES

We are often interested in charting thermodynamic processes between states on thermodynamic coordinates. A state only when a system is in equilibrium. If a process involves finite, unbalanced forces, the system can pass through non-equilibrium states, which we cannot treat. An extremely useful idealization, however, is that only "infinitesimal" unbalanced forces exist, so that the process can be viewed as taking place in a series of "quasi-equilibrium" states. (The term quasi can be taken to mean "as if;" it used in a number of contexts such as quasi-one-dimensional, quasi-steady, etc.) For this to be true the process must be slow in relation to the time needed for the system to come to equilibrium internally.

For a gas at conditions of interest to us, a given molecule can undergo roughly 10^{10} molecular collisions per second, so that, if ten collisions are needed to come to equilibrium, the equilibration time is on the order of 10^{-9} seconds. This is generally much shorter than the time scales associated with the bulk properties of the flow (say the time needed for a fluid particle to move some significant fraction of the length of the device of interest). Over a large range of parameters, therefore, it is a very good approximation to view the thermodynamic processes as consisting of such a succession of equilibrium states, which we can chart.

The figures below demonstrate the use of thermodynamics coordinates to plot isolines, lines along which a property is constant. They include constant temperature lines, or isotherms, on a p-v diagram, constant volume lines, or isochors on a T-p diagram, and constant pressure lines, or isobars, on a T-v diagram for an ideal gas.

Real substances may have phase changes which we can also plot on thermodynamic coordinates.

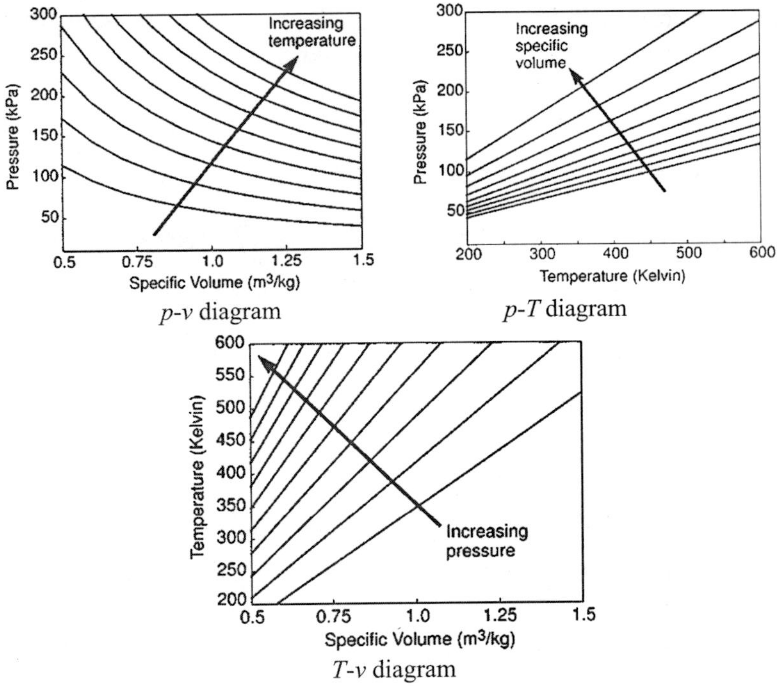

Fig. Thermodynamics Coordinates and Isolines for an Ideal Gas

EQUATIONS OF STATE

It is an experimental fact that two properties are needed to define the state of any pure substance in equilibrium or undergoing a steady or quasi-steady process. Thus for a simple compressible gas like air,

$$P = P(v, T), \text{ or } v = v(P, T), \text{ or } T = T(P, v),$$

where v is the volume per unit mass, $1/\rho$. In words, if we know v and T we know P, etc.

Any of these is equivalent to an equation $f(P, v, T) = 0$, which is known as an equation of state. The equation of state for an ideal gas, which is a very good approximation to real gases at conditions that are typically of interest for aerospace applications is

$$P\bar{v} = RT,$$

where $\bar{v}$ is the volume per mol of gas and $\mathcal{R}$ is the "Universal Gas Constant," 8.31 kJ/kmol-K.

A form of this equation which is more useful in fluid flow problems is obtained if we divide by the molecular weight, $\mathcal{M}$:

$$Pv = RT, \ or \quad P = \rho RT$$

where is $\mathcal{R}/\mathcal{M}$, which has a different value for different gases due to the different molecular weights. For air at room conditions, R = 0.287 kJ/kg–K.

CHANGING THE STATE OF A SYSTEM WITH HEAT AND WORK

Changes in the state of a system are produced by interactions with the environment through heat and work, which are two different modes of energy transfer. During these interactions, equilibrium (a static or quasi-static process) is necessary for the equations that relate system properties to one-another to be valid.

HEAT

Heat is energy transferred due to temperature differences only:

- Heat transfer can alter system states;
- Bodies don't "contain" heat; heat is identified as it comes across system boundaries;
- The amount of heat needed to go from one state to another is path dependent;
- Adiabatic processes are ones in which no heat is transferred.

ZEROTH LAW OF THERMODYNAMICS

With the material so far, we are now in a position to describe the Zeroth Law. Like the other laws of thermodynamics the Zeroth Law is based on observation. We start with two such observations:

- If two bodies are in contact through a thermally-conducting boundary for a sufficiently long time, no further observable changes take place; thermal equilibrium is said to prevail.
- Two systems which are individually in thermal equilibrium with a third are in thermal equilibrium with each other; all three systems have the same value of the property called temperature.

These closely connected ideas of temperature and thermal

equilibrium are expressed formally in the "Zeroth Law of Thermodynamics"

Zeroth Law: There exists for every thermodynamic system in equilibrium a property called temperature. Equality of temperature is a necessary and sufficient condition for thermal equilibrium. The Zeroth Law thus defines a property (temperature) and describes its behaviour.

Note that this law is true regardless of how we measure the property temperature. (Other relationships we work with will typically require an absolute scale, so in these notes we use either the Kelvin K = 273.15 + °C or Rankine scales) The zeroth law is depicted schematically.

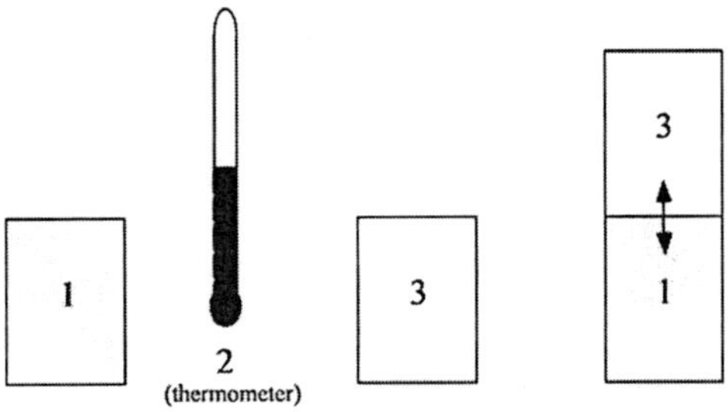

if $T_1 = T_2$ and $T_2 = T_3$ then ${}_1Q_3 = 0$

Fig. The Zeroth Law Schematically

WORK

Heat is a way of changing the energy of a system by virtue of a temperature difference only. Any other means for changing the energy of a system is called work. We can have push-pull work (e.g. in a piston-cylinder, lifting a weight), electric and magnetic work (e.g. an electric motor), chemical work, surface tension work, elastic work, etc. In defining work, we focus on the effects that the system (e.g. an engine) has on its surroundings. Thus we define work as being positive when the system does work on the surroundings (energy leaves the system). If work is done on the system (energy added to the system), the work is negative.

Consider a simple compressible substance, a gas (the system), exerting a force on the surroundings via a piston, which moves through some distance, l. The work done on the surroundings, W_{onsurr}, is

$$dW_{\text{on surr.}} = \text{Force on surr.} \times \text{dl}$$

$$dW_{\text{on surr.}} = \frac{\text{Force on surr.}}{\text{Area}} \times (\text{Area} \times dl)$$

$$dW_{\text{on surr.}} = \text{pressure of surr.} \times d\,\text{Volume}$$

$$dW_{\text{on surr.}} = p_x \times dV$$

therefore

$$W_{\text{on surr.}} = \int_{V_1}^{V_2} p_x dV.$$

Why is the pressure p_x instead of p_s? Consider $p_x = 0$ (vacuum). No work is done on the surroundings even though p_s changes and the system volume changes.

Use of p_x instead of p_s is often inconvenient because it is usually the state of the system that we are interested in. The external pressure can only be related to the system pressure if $p_x \approx p_x$. For this to occur, there cannot be any friction, and the process must also be slow enough so that pressure differences due to accelerations are not significant. In other words, we require a "quasi-static" process, $p_x \approx p_x$. Consider $p_x = p_s \pm dp$

$$\text{W} = \int_{V_1}^{V_2} p_x dV = \int_{V_1}^{V_2} (p_s \pm dp) dV = \int_{V_1}^{V_2} p_s dV \pm dpdV.$$

Therefore, when *dp* is small (the process is quasi-static),

$$\text{W} = \int_{V_1}^{V_2} p_s dV,$$

and the work done by the system is the same as the work done on the surroundings

Under these conditions, we say that the process is "reversible." The conditions for reversibility are that:

- If the process is reversed, the system and the surroundings will be returned to the original states.
- To reverse the process we need to apply only an infinitesimal *dp*. A reversible process can be altered in direction by infinitesimal changes in the external conditions.

Remember this result, that we can only relate work done on surroundings to system pressure for quasi-static (or reversible) processes. In the case of a "free expansion," where $p_x = 0$ (vacuum), p_s is not related to p_x (and thus, not related to the work) because the system is not in equilibrium.

We can write the above expression for work done by the system in terms of the specific volume, v,

$$W = m\int_{v_1}^{v_2} p_s dv.$$

where is the mass of the system. Note that if the system volume expands against a force, work is done by the system. If the system volume contracts under a force, work is done on the system.

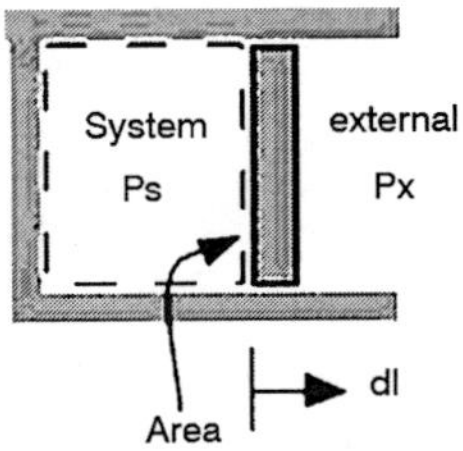

Fig. A Closed System (Dashed Box) Against a Piston, which Moves Into the Surroundings

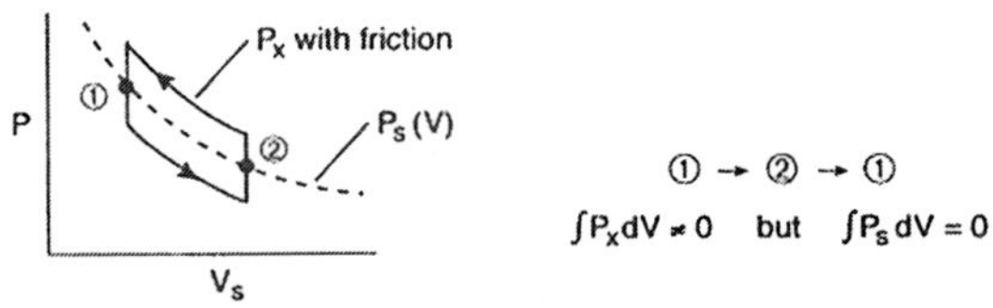

Fig. Work During an Irreversible Process $\neq \int P_s dV$.

For simple compressible substances in reversible processes, the work done can be represented as the area under a curve in a pressure-volume diagram.

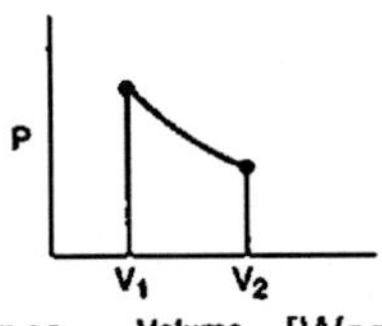

[Work is area under curve of $P(V)$] [Work depends on path]

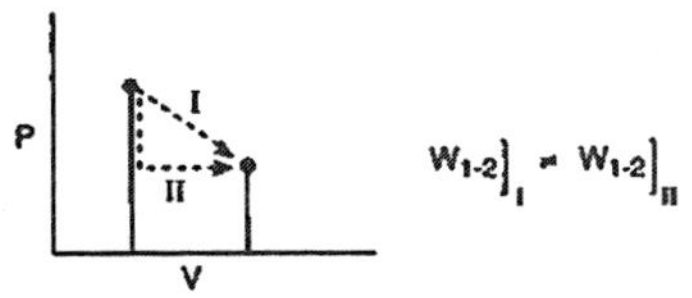

Fig. Work in P - V coordinates

Key points to note are the following:

- Properties only depend on states, but work is path dependent (depends on the path taken between states); therefore work is not a property, and not a state variable.
- When we say W_{1-2} , the work between states 1 and 2, we need to specify the path.
- For irreversible (non-reversible) processes, we cannot use $\int pdV$; either the work must be given or it must be found by another method.

Example: Work on Two Simple Paths

Consider Figure , which shows a system undergoing quasi-static processes for which we can calculate work interactions as $\int pdV$.

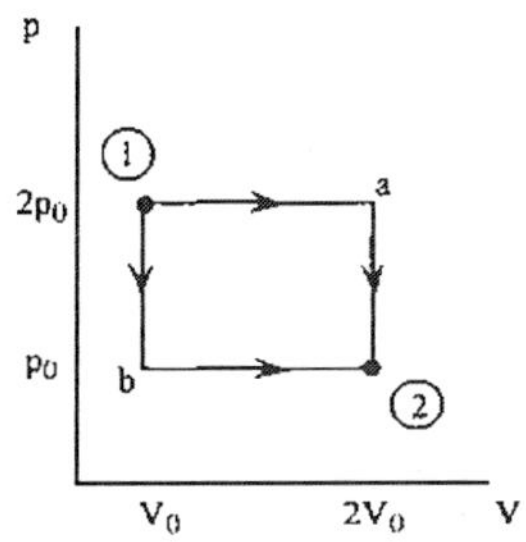

Fig. Simple processes

$$W = 2p_0\,(2V_0 - V_0) = 2p_0V_0$$

Along Path a:

$$W = p_0\,(2V_0 - V_0) = p_0V_0$$

Along Path b:

Practice Questions

Given a piston filled with air, ice, a bunsen burner, and a stack of small weights, describe

- how you would use these to move along either path a or path b above, and
- how you would physically know the work is different along each path.

Example: Work Done During Expansion of a Gas

Consider the quasi-static, isothermal expansion of a thermally ideal gas from p_1, V_1 to p_2, V_2. To find the work we must know the path. Is it specified? Yes, the path is specified as isothermal.

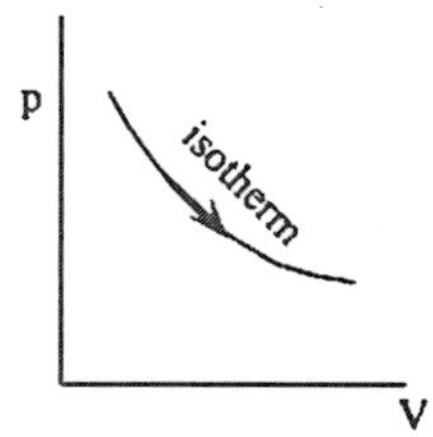

Fig. Quasi-static, Isothermal Expansion of an Ideal Gas

The equation of state for a thermally ideal gas is

$$pV = n\mathcal{R}T,$$

where n is the number of moles, $\mathcal{R}$ is the Universal gas constant, and V is the total system volume. We write the work as above, substituting the ideal gas equation of state,

$$W = \int_{V_1}^{V_2} \frac{n\mathcal{R}T}{V} dV = = n\mathcal{R}T \int_{V_1}^{V_2} \frac{dV}{V} = n\mathcal{R}T \ln\left(\frac{V_2}{V_1}\right)$$

also for T, = constant, $p_1V_1 = p_2\ V_2$, so the work done by the system is

$$W = n\mathcal{R}T \ln\left(\frac{V_2}{V_1}\right) = n\mathcal{R}T \ln\left(\frac{p_1}{p_2}\right)$$

or in terms of the specific volume and the system mass,

$$W = mRT \ln\left(\frac{v_2}{v_1}\right) = mRT \ln\left(\frac{p_1}{p_2}\right).$$

Work vs. Heat

We can have one, the other, or both: it depends on what crosses the system boundary (and thus, on how we define our system). Consider a resistor that is heating a volume of water:

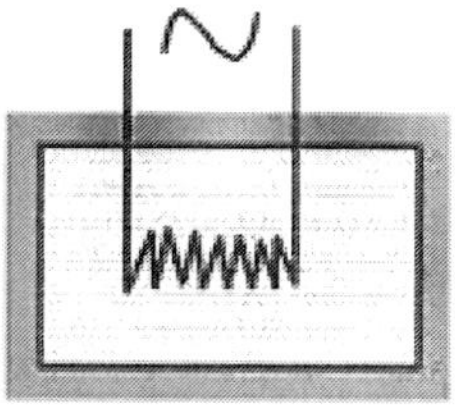

Fig. A resistor Heating Water

- If the water is the system, then the state of the system will be changed by heat transferred from the resistor.

- If the system is the water and the resistor combined, then the state of the system will be changed by electrical work.

CONSERVATION OF ENERGY

FIRST LAW OF THERMODYNAMICS

Observation leads to the following two assertions:

- There exists for every system a property called energy, *E*. The system energy can be considered as a sum of internal energy, kinetic energy, potential energy, and chemical energy.
  - Like the Zeroth Law, which defined a useful property, "temperature," the First Law defines a useful property called "energy."
  - The two new terms (compared to what you have seen in physics and dynamics) are the internal energy and the chemical energy. For most situations in this class, we will neglect the chemical energy. We will generally not, however, neglect the internal energy, . It arises from the random or disorganized motion of molecules in the system. Since this molecular motion is primarily

a function of temperature, the internal energy is sometimes called "thermal energy."

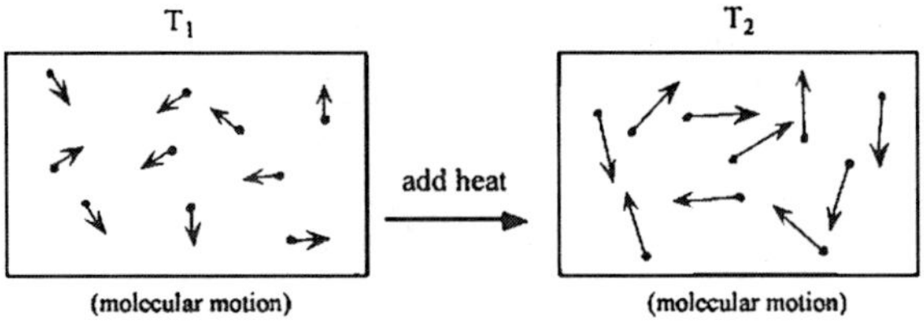

Fig. Random Motion is the Physical Basis for Internal Energy

The internal energy, u, is a function of the state of the system. Thus $u = u\ (p, T)$, or $u = u\ (p, v)$, or $u = u\ (p, T)$. Recall that for pure substances the entire state of the system is specified if any two properties are specified.

- The change in energy of a system is equal to the difference between the heat added to the system and the work done by the system,

$\Delta E = Q - W$ (units are Joules, J),

where E is the energy of the system,

Q is the heat input to the system, and

W is the work done by the system.

$E = U$ (thermal energy) $+ E_{kinctic} + E_{potential} + \ldots$

- Like the Zeroth Law, the First Law describes the behaviour of the new property.
- The equation can also be written on a per unit mass basis

$$\Delta e = q - w \text{ (units are J/kg).}$$

- In many situations the potential energy, kinetic energy, and chemical energy of the system are constant or not important. Then

$$\Delta E = \Delta U,$$

and

$$\Delta U = Q - W \text{ or } \Delta u = q - w.$$

- Note that Q and W are not functions of state, but U, which arises from molecular motion, depends only on the state of the system; does not depend on how the system got to that state. We therefore have the striking result that:

ΔU is independent of path even though Q and W are not!

Sometimes this difference is emphasized by writing the First Law in differential form,

$$dU = \delta Q - \delta W \quad \text{or} \quad du = \delta q - \delta w,$$

where the symbol "δ" is used to denote that these are not exact differentials but rather are dependent on path.

- *Note that the signs are important:* Q is defined to be positive if it is transferred to the system; thus the numerical value we substitute for Q will be positive if heat is transferred to the system from the surroundings, and negative if heat is transferred from the system to the surroundings.
  W is defined to be positive if it is done by the system; thus the numerical value we substitute for W will be positive if the system is doing work, and negative if work is being done on the system.
- For quasi-static processes we can substitute $W = p_{sys}dV$,

$$dU = dQ - pdV \quad \text{or} \quad du = \delta q - pdv$$

To give an example of where the first law is applied. We heat a gas, it expands against a weight, some force (pressure times area) is applied over a distance, and work is done. The change in energy of the system supplies the connection between the heat added and work done. We will spend most of the course dealing with various applications of the first law — in one form or another.

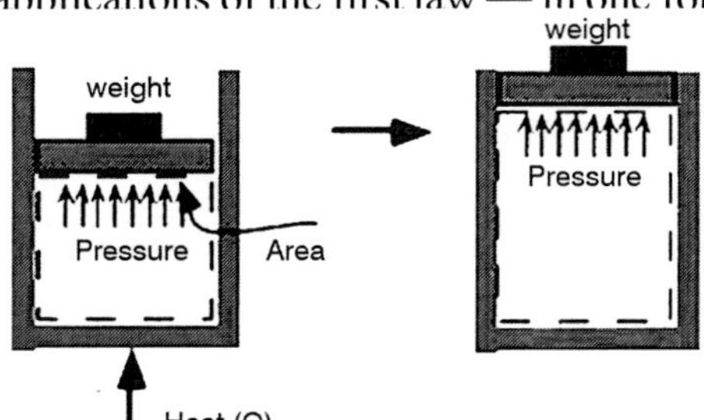

Fig. The Change in Energy of a System Relates the Heat Added to the Work Done

The form of the first law we have given here is sometimes called the "control mass" form, because it is well suited to dealing with systems of a fixed mass.

COROLLARIES OF THE FIRST LAW

- Work done in any adiabatic ($Q = 0$) process is a function of state. We can write the first law, setting the heat transfer term equal to zero, as

$$\Delta U = \cancel{Q}^{0} - W.$$

Since ΔU depends only on the state change, now W can be found as a function of the state change.

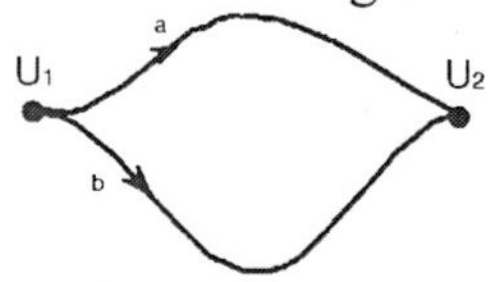

Fig. The change in energy between two states is not path dependent.

- For a cyclic process heat and work transfers are numerically equal

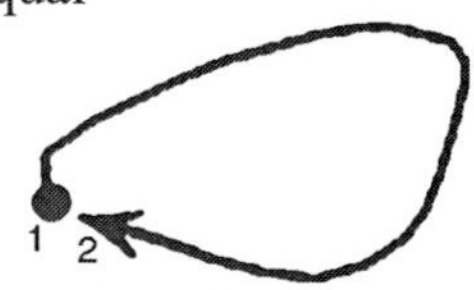

Fig. Since Energy is a Function of State Only, any Process that Returns a System to its Original State Leaves its Energy Unchanged.

$$U_{\text{final}} = U_{\text{initial}}$$

therefore

$$\Delta U = 0$$

and

$$Q = W \quad \text{or} \quad \oint \delta Q = \oint \delta W$$

APPLICATIONS OF THE FIRST LAW

ADIABATIC, STEADY, THROTTLING OF A GAS

The configuration of interest. We wish to know the relation between properties upstream of the valve, denoted by "1" and those downstream, denoted by "2".

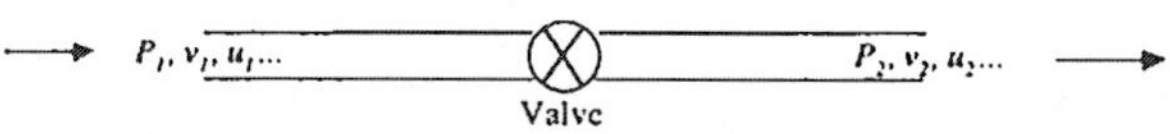

Fig. Adiabatic flow Through a Valve, a Generic Throttling Process

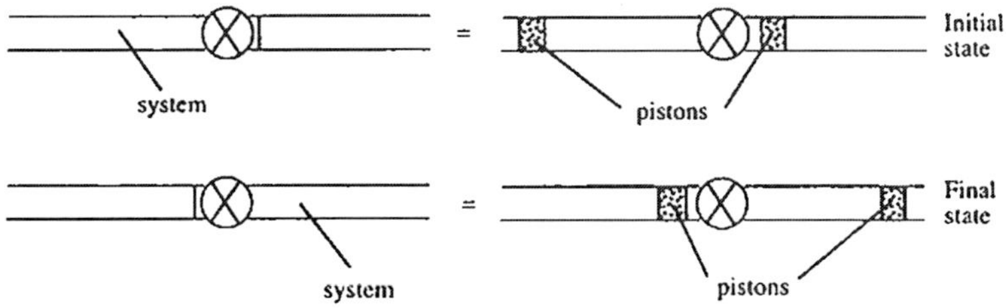

Fig. Equivalence of Actual System and Piston Model

To analyze this situation, we can define the system (choosing the appropriate system is often a critical element in effective problem solving) as a unit mass of gas in the following two states. Initially the gas is upstream of the valve and just through the valve. In the final state the gas is downstream of the valve plus just before the valve.

In terms of the system behaviour, however, we could replace the fluid external to the system by pistons which exert the same pressure that the external fluid exerts, as indicated schematically on the right side.

The process is adiabatic, with changes in potential energy and kinetic energy assumed to be negligible. The first law for the system is therefore

$$\Delta U = -W.$$

The work done by the system is

$$W = P_2V_2 - P_1V_1.$$

Use of the first law leads to

$$U_2 = P_2V_2 = U_1 + P_1V_1.$$

In words, the initial and final states of the system have the same value of the quantity $U + PV$. For the case examined, since we are dealing with a unit mass, the initial and final states of the system have the same value of $u + Pv$.

We define this quantity as the "enthalpy," usually denoted by H,

$$H = U + PV.$$

In terms of the specific quantities, the enthalpy per unit mass is

$$h = u + Pv = u + P/\rho.$$

It is a function of the state of the system. H has units of Joules, and has units of Joules per kilogram.

The utility and physical significance of enthalpy will become clearer as we work with more flow problems. When you evaluate the energy of an object of volume V, you have to remember that the object had to push the surroundings out of the way to make room for itself. With pressure p on the object, the work required to make a place for itself is pV. This is so with any object or system, and this work may not be negligible. The force of one atmosphere pressure on one square meter is equivalent to the force of a mass of about 10 tons. Thus the total energy of a body is its internal energy plus the extra energy it is credited with by having a volume V at pressure p. We call this total energy the enthalpy, H.

QUASI-STATIC EXPANSION OF A GAS

Consider a quasi-static process of constant pressure expansion. We can write the first law in terms of the states before and after the expansion as

$$Q = (U_2 - U_1) + W,$$

and writing the work in terms of system properties,

$$= (U_2 - U_1) + p(V_2 - V_1) \quad \text{since } p_1 = p_2 = p.$$

By grouping terms we can write the heat input in terms of the enthalpy change of the system:

$$Q = (U_2 + pV_2) - (U_1 + pV_1)$$
$$= H_2 - H_1$$

TRANSIENT FILLING OF A TANK

Another example of a flow process, this time for an unsteady flow, is the transient process of filling a tank, initially evacuated, from a surrounding atmosphere, which is at a pressure P_0 and a temperature T_0.

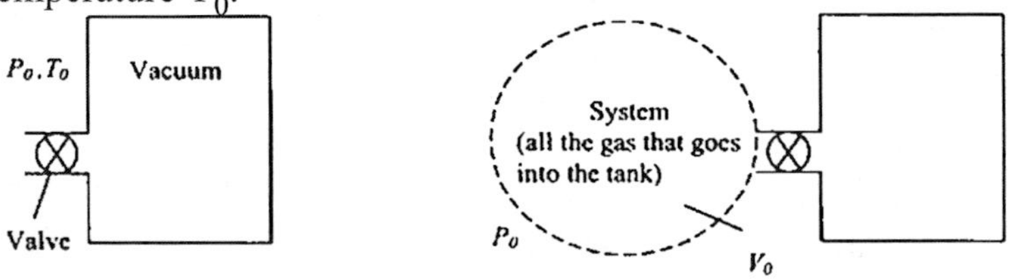

Fig. A Transient Problem-filling of a Tank from the Atmosphere

At a given time, the valve at the tank inlet is opened and the outside air rushes in. The inflow stops when the pressure inside is equal to the pressure outside. The tank is insulated, so there is no heat transfer to the atmosphere. What is the final temperature of the gas in the tank?

This time we take the system to be all the gas that enters the tank. The initial state has the system completely outside the tank, and the final state has the system completely inside the tank. The kinetic energy initially and in the final state is negligible, as is the change in potential energy, so the first law again takes the form

$$\Delta U = -W.$$

Work is done on the system, of magnitude $P_0 V_0$, where V_0 is the initial volume of the system, so

$$\Delta U = P_0\, V_0.$$

In terms of quantities per unit mass ($\Delta U = m\Delta u$, $V_0 = mv_0$, where m is the mass of the system),

$$\Delta u = u_{\text{final}} - u_i = P_0 v_0.$$

The final value of the internal energy is

$$\begin{aligned} u_{\text{final}} &= u_i + P_0 v_0 \\ &= h_i = h_0. \end{aligned}$$

For a perfect gas with constant specific heats,

$$u = c_v T, \quad h = c_p T,$$

$$c_v T_{\text{final}} = c_p T_0,$$

$$T_{\text{final}} = \frac{c_p}{c_v} T_0 = \gamma T_0.$$

The final temperature is thus roughly 200°F hotter than the outside air!

It may be helpful to recap what we used to solve this problem. There were basically four steps:

- Definition of the system
- Use of the first law
- Equating the work to a "PdV" term
- Assuming the fluid to be a perfect gas with constant specific heats.

A message that can be taken from both of these examples (as well as from a large number of other more complex situations, is that the quantity $h = u + Pv$ occurs naturally in problems of fluid flow. Because the combination appears so frequently, it is not only defined but also tabulated as a function of temperature and pressure for a number of working fluids.

THE FIRST LAW IN TERMS OF ENTHALPY

We start with the first law in differential form and substitute pdV for dW by assuming a quasi-static or reversible process:

$dU = \delta Q - \delta W$ (true forany process, neglecting ΔKE and ΔPE)

$dU = \delta Q - pdV$ (true for any quasi-static process, no ΔKE or ΔPE)

The definition of enthalpy,

$$H = U + pV,$$

can be differentiated (applying the chain rule to the pV term) to produce

$$dH = dU + pdV + Vdp$$

Substituting the dU above for the dU in the First Law, we obtain

$dH = \delta Q - \delta W + pdV + Vdp$ (valid for any process)

or

$dH = \delta Q + Vdp$ (valid for any quasi-static process)

THE RELATION BETWEEN TEMPERATURE CHANGE AND HEAT

How much does a given amount of heat transfer change the temperature of a substance? It depends on the substance. In general

$$Q = C\Delta T,$$

where C is a constant that depends on the substance. We can determine the constant for any substance if we know how much heat is transferred. Since heat is path dependent, however, we must specify the process, i.e., the path, to find C.

Two useful processes are constant pressure and constant volume, so we will consider these each in turn. We will call the specific heat at constant pressure Cp, and that at constant volume Cv, or cp and cv per unit mass.

- The Specific Heat at Constant Volume

Remember that if we specify any two properties of the system, then the state of the system is fully specified. In other words we can write $u = u\ (T, v)$ $u = u\ (p, v)$ or $u = u\ (p, T)$.

Consider the form $u = u\ (T, v)$, and use the chain rule to write how changes with respect to T and v:

$$du = \left(\frac{\partial u}{\partial T}\right)_v dT + \left(\frac{\partial u}{\partial v}\right)_T dv.$$

For a constant volume process, the second term is zero since there is no change in volume, $dv = 0$. Now if we write the First Law for a quasi-static process, with $dW = pdv$,

$$du = \delta q - pdv,$$

we see that again the second term is zero if the process is also constant volume. Equating Equation with dv canceled in each,

$$\delta q = \left(\frac{\partial u}{\partial T}\right)_v dT,$$

and rearranging

$$\left(\frac{\partial u}{\partial T}\right)_v = \left(\frac{\partial q}{\partial T}\right)_v.$$

In this case, any energy increase is due only to energy transfer as heat. We can therefore use our definition of specific heat from Equation to define the specific heat for a constant volume process,

$$c_v \equiv \left(\frac{\partial u}{\partial T}\right)_v.$$

- The Specific Heat at Constant Pressure

If we write $h = h\ (T, p)$, and consider a constant pressure process, we can perform a similar derivation to the one above and show that

$$c_p \equiv \left(\frac{\partial h}{\partial T}\right)_p.$$

In the derivation of c_v, we considered only a constant volume process, hence the name, "specific heat at constant volume." It is more useful, however, to think of c_n in terms of its definition as a certain partial derivative, which is a thermodynamic property, rather than as a quantity related to heat transfer in a special process. In fact, the derivatives above are defined at any point in any quasi-static process whether that process is constant volume, constant pressure, or neither.

The names "specific heat at constant volume" and "specific heat at constant pressure" are therefore unfortunate misnomers; c_v and c_p are thermodynamic properties of a substance, and by definition depend only the state. They are extremely important values, and have been experimentally determined as a function of the thermodynamic state for an enormous number of simple compressible substances.

To recap:

$$c_p = \left(\frac{\partial h}{\partial T}\right)_p \quad \text{and} \quad c_v = \left(\frac{\partial u}{\partial T}\right)_v .$$

or

$$C_p = \left(\frac{\partial H}{\partial T}\right)_p \quad \text{and} \quad C_v = \left(\frac{\partial U}{\partial T}\right)_v .$$

Practice Questions Throw an object from the top tier of the lecture hall to the front of the room. Estimate how much the temperature of the room has changed as a result. Start by listing what information you need to solve this problem.

SPECIFIC HEATS OF AN IDEAL GAS

The equation of state for an ideal gas is

$$PV = NRT$$

where N is the number of moles of gas in the volume V. Ideal gas behaviour furnishes an extremely good approximation to the behaviour of real gases for a wide variety of aerospace applications. It should be remembered, however, that describing a substance as an ideal gas constitutes a model of the actual physical situation, and the limits of model validity must always be kept in mind.

One of the other important features of an ideal gas is that its internal energy depends only upon its temperature. (For now, this can be regarded as another aspect of the model of actual systems that the ideal gas represents, but it can be shown that this is a consequence of the form of the equation of state.) Since u depends only on T,

$$du = \left(\frac{\partial u}{\partial T}\right)_v dT + \left(\frac{\partial u}{\partial v}\right)_T^{\nearrow 0} dv$$

or

$$du = cv\ (T)dT.$$

In the above equation we have indicated that c_v can depend on T. Like the internal energy, the enthalpy is also only dependent on T for an ideal gas. (If u is a function of T, then, using the ideal gas equation of state, $u + Pv$ is also.) Therefore,

$$dh = \left(\frac{\partial h}{\partial T}\right)_v dT + \left(\frac{\partial h}{\partial p}\right)_T^{\nearrow 0} dp,$$

and

$$dh = c_p\ (T)dT.$$

If we are interested in finite changes of internal energy or enthalpy, we integrate,

$$\Delta u_{1-2} = \int_{T_1}^{T_2} c_v(T)dT$$

and

$$\Delta h_{1-2} = \int_{T_1}^{T_2} c_p(T)dT$$

Over small temperature changes (DT ≈ 200), it is often assumed that c_v and c_p are constant. Furthermore, there are wide ranges over which specific heats do not vary greatly with respect to temperature. It is thus often useful to treat them as constant. If so

$$du = c_v dT,$$

$$dh = c_p dT,$$

$$u_2 - u_1 = c_v(T_2 - T_1),$$

$$u_2 - h_1 = c_p(T_2 - T_1),$$

These equations are useful in calculating internal energy or enthalpy differences, but it should be remembered that they hold only if the specific heats are constant. We can relate the specific heats of an ideal gas to its gas constant as follows. We write the first law in terms of internal energy,

$$\delta q = du + dW,$$

and assume a quasi-static process so that we can also write it in terms of enthalpy,

$$\delta q = dh - vdp.$$

Equating the two first law expressions above, and assuming an ideal gas, we obtain

$$c_p dT - vdp = c_v dT + pdv.$$

Combining terms,

$$(c_p - c_v)\, dT = d(pv)$$

$$c_p - c_v = \frac{d(pv)}{dT}.$$

Since $pv = RT$,

$$c_p - c_v = R$$

An expression that will appear often is the ratio of specific heats, which we will define as

$$\gamma = \frac{c_p}{c_v}.$$

Below we summarize the important results for all ideal gases, and give some values for specific types of ideal gases.

- All ideal gases:
- The specific heat at constant volume (c_v for a unit mass or Cv for one kmol) is a function of T only.

- The specific heat at constant pressure (c_p for a unit mass or Cp for one kmol) is a function of T only.
- A relation that connects the specific heats c_p, c_v, and the gas constant is

$$C_P - C_V = R,$$

where the units depend on the mass considered. For a unit mass of gas, e.g., a kilogram, c_p and c_v would be the specific heats for one kilogram of gas and R is as defined above. For one kmol of gas, the expression takes the form

$$C_p - C_v = R,$$

where C_P and C_V have been used to de note the specific heats for one kmol of gas and R is the universal gas constant.

- The specific heat ratio, $\gamma = c_p/c_v$ (or C_P/C_V), is a function of T only and is greater than unity
- An ideal gas with specific heats independent of temperature, c_p = const and c_v = const, is referred to as a perfect gas. Monatomic gases and diatomic gases at ordinary temperatures are considered perfect gases. To make this distinction the terminology "a perfect gas with constant specific heats" is used throughout the notes. In some textbooks perfect gases are sometimes also referred to as ideal gases, and to avoid confusion we use the stated terminology.
- Monatomic gases, such as He, Ne, Ar, and most metallic vapors:
- c_v (or C_V) is constant over a wide temperature range and is very nearly equal to (3/2)R[or (3/2)R, for one kmol].
- c_p (or C_P) is constant over a wide temperature range and is very nearly equal to (5/2)R[or (5/2)R, for one kmol].
- γ is constant over a wide temperature range and is very nearly equal to 5/3 [γ = 1.67].
- So-called permanent diatomic gases, namely H_2, O_2, N_2, Air, NO, and CO:
- c_v (or C_V) is nearly constant at ordinary temperatures, being approximately (5/2)R [(5/2)R, for one kmol], and increases slowly at higher temperatures.
- c_p (or C_P) is nearly constant at ordinary temperatures, being approximately (7/2) R[(7/2)R, for one kmol], and increases slowly at higher temperatures.
- γ is constant over a temperature range of roughly to and is very nearly equal to 7/5[γ = 1.4]. It decreases with temperature above this.
- Polyatomic gases and gases that are chemically active, such as CO_2, NH_3, CH_4, and Freons:

The specific heats, c_v and c_p, and g vary with the temperature, the variation being different for each gas. The general trend is that heavy molecular weight gases (i.e., more complex gas molecules than those listed in 2 or 3), have values of g closer to unity than diatomic gases, which, are closer to unity than monatomic gases. Values of g below 1.2 are typical of Freons which have molecular weights of over one hundred.

In general, for substances other than ideal gases, u and h depend on pressure as well as on temperature, and the above relations will not all apply. In this respect, the ideal gas is a very special model.

In summary, the specific heats are thermodynamic properties and can be used even if the processes are not constant pressure or constant volume. The simple relations between changes in energy (or enthalpy) and temperature are a consequence of the behaviour of an ideal gas, specifically the dependence of the energy and enthalpy on temperature only, and are not true for more complex substances.

REVERSIBLE ADIABATIC PROCESSES FOR AN IDEAL GAS

From the first law, with $Q = 0$ $du = c_v dT$, and Wor $= pdv$,

$$du + pdv = 0$$

Also, using the definition of enthalpy,

$$dh = \underline{du + pdv} + vdp.$$

The underlined terms are zero for an adiabatic process. Rewriting equations,

$$\gamma c_v dT = -\,\gamma pdv$$

$$c_v dT = vdp.$$

Combining the above two equations we obtain

$$-\gamma pdv = vdp \quad \text{or} \quad -\gamma dv / v = dp / p.$$

Equation can be integrated between states 1 and 2 to give

$$-\gamma / n(v_2 / v_1) = \ln (p_2 / p_1), \quad \text{or equivalently},$$

$$(p_2 v_2^{\gamma})/(p_1 v_1^{\gamma}) = 1.$$

For an ideal gas undergoing a reversible, adiabatic process, the relation between pressure and volume is thus:

$$pv^{\gamma} = \text{constant, or}$$

$$p = \text{constant} \times \rho^{\gamma}.$$

We can substitute for p or v in the above result using the ideal gas law, or carry out the derivation slightly differently, to also show that

$$\frac{p_2}{p_1} = \left(\frac{T_2}{T_1}\right)^{\frac{\gamma}{\gamma-1}} \quad \text{and} \quad \frac{T_2}{T_1} = \left(\frac{v_1}{v_2}\right)^{\gamma-1}$$

We will use the above equations to relate pressure and temperature to one another for quasi-static adiabatic processes (for instance, this type of process is our idealization of what happens in compressors and turbines).

CONTROL VOLUME FORM OF THE SYSTEM LAWS

The thermodynamic laws (as well as Newton's laws) are for a system, a specific quantity of matter. More often, in propulsion and power problems, we are interested in what happens in a fixed volume, a rocket motor or a jet engine through which mass is flowing at a certain rate. We may also be interested in the rates of heat and work into and out of a system.

For this reason, the control volume form of the system laws is of great importance. Rather than focus on a particle of mass which moves through the engine, it is more convenient to focus on the volume occupied by the engine. This requires us to use the control volume form of the thermodynamic laws, developed below.

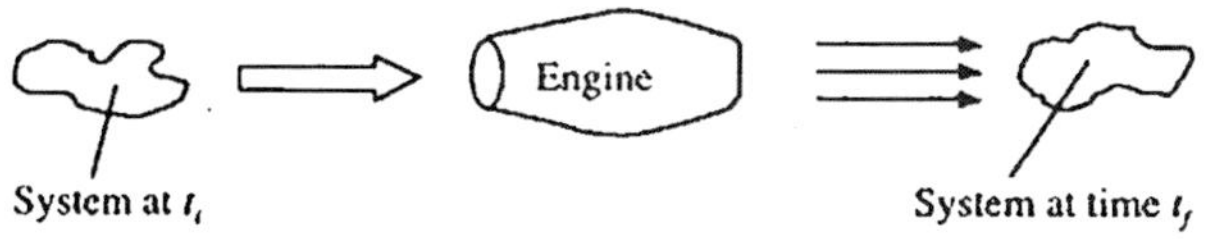

Fig. Control Volume and System for Flow Through a Propulsion Device

CONSERVATION OF MASS

For the control volume shown, the rate of change of mass inside the volume is given by the difference between the mass flow rate in and the mass flow rate out. For a single flow coming in and a single flow coming out this is

$$\frac{dm_{cv}}{dt} = \dot{m}_{\text{in}} - \dot{m}_{\text{out}}$$

If the mass inside the control volume changes with time it is because some mass is added or some is taken out. In the special case of a steady flow, $d/dt = 0$, therefore

$$\dot{m}_{\text{in}} = \dot{m}_{\text{out}} = \dot{m}.$$

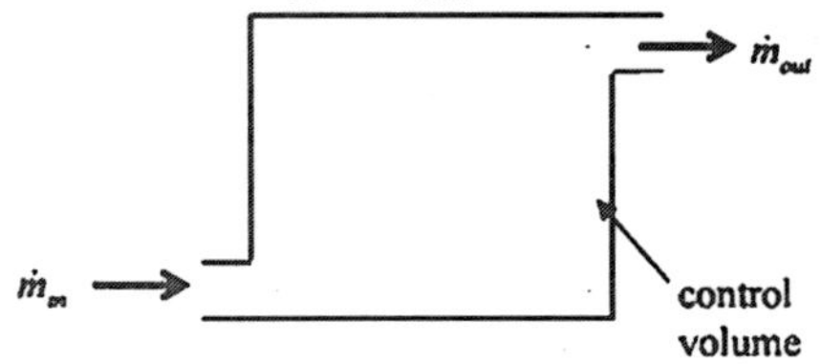

Fig. A Control Volume used to Track Mass Flows

CONSERVATION OF ENERGY

The first law of thermodynamics can be written as a rate equation:

$$\frac{dE}{dt} = \dot{Q} = \dot{W},$$

where

$$\dot{Q} = \lim_{dt \to 0} \left(\frac{\delta Q}{dt} \right) \text{ rate of total heat transfer to the system}$$

$$\dot{W} = \lim_{dt \to 0} \left(\frac{\delta W}{dt} \right) \text{ rate of total work done by the system.}$$

To derive the first law as a rate equation for a control volume we proceed as with the mass conservation equation. The physical idea is that any rate of change of energy in the control volume must be caused by the rates of energy flow into or out of the volume.

The heat transfer and the work are already included and the only other contribution must be associated with the mass flow in and out, which carries energy with it.

The desired form of the equation will be (rate of change of energy in cv) = (rate of heat added to C.V.) – (rate of work done) + (rate of energy flow into C.V.) – (rate of energy flow out C.V.).

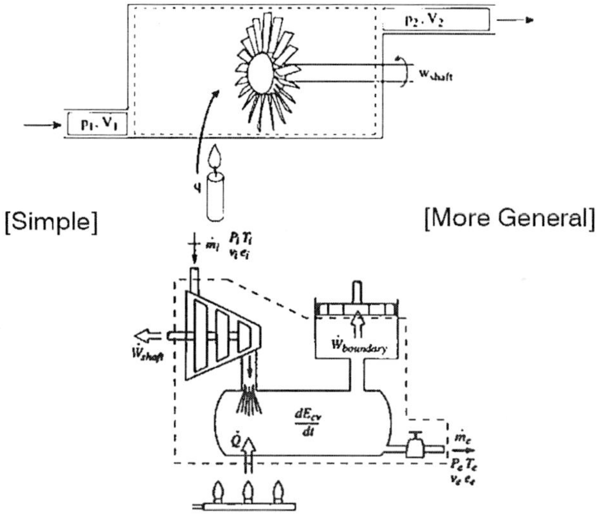

Fig. Schematic Diagrams Illustrating Terms in the Energy Equation for a Simple and a more General Control Volume

The fluid that enters or leaves has an amount of energy per unit mass given by

$$e = u + c^2/2 + gz,$$

where c is the fluid velocity relative to some coordinate system, and we have neglected chemical energy. In addition, whenever fluid enters or leaves a control volume there is a work term associated with the entry or exit. The present derivation is essentially an application of the ideas presented there. Flow exiting at station "e" must push back the surrounding fluid, doing work on it. Flow entering the volume at station "*i*" is pushed on by, and receives work from the surrounding air.

The rate of flow work at exit is given by the product of the pressure times the exit area times the rate at which the external flow is "pushed back." The latter, however, is equal to the volume per unit mass times the rate of mass flow. Put another way, in a time dt, the work done on the surroundings by the flow at the exit station is

$$dW_{\text{flow}} = pvdm_c.$$

The net rate of flow work is

$$\dot{W}_{\text{flow}} = p_c v_c \dot{m}_c - p_i v_i \dot{m}_i.$$

Including all possible energy flows (heat, shaft work, shear work, piston work, etc.), the first law can then be written as:

$$\frac{d}{dt}\sum E_{cv} = \sum \dot{Q}_{cv} + \sum \dot{W}_{\text{shaft}} + \sum \dot{W}_{\text{shcar}} + \sum \dot{W}_{\text{piston}} + \sum \dot{W}_{\text{flow}} + \sum \dot{m}\left(u + \frac{c^2}{2} + gz\right)$$

where Σ includes the sign associated with the energy flow. If heat is added or work is done on the system then the sign is positive, if work or heat are extracted from the system then the sign is negative. NOTE: this is consistent with $\Delta E = Q - W$, where W is the work done by the system on the environment, thus work is flowing out of the system.

We can then combine the specific internal energy term, u, in e and the specific flow work term, pv, to make the enthalpy appear:

Total energy associated with mass flow:

$$e + pv = u + c^2/2 + gz + pv = h + c^2/2 + gz.$$

Thus, the first law can be written as:

$$\frac{d}{dt}\sum E_{cv} = \sum \dot{Q}_{cv} + \sum \dot{W}_{\text{shaft}} + \sum \dot{W}_{\text{shcar}} + \sum \dot{W}_{\text{piston}} + \sum \dot{m}\left(h + \frac{c^2}{2} + gz\right).$$

For most of the applications in this course, there will be no shear work and no piston work. Hence, the first law for a control volume will be most often used as:

$$\frac{dE_{cv}}{dt} = \dot{Q}_{cv} - \dot{W}_{\text{shaft}} + \dot{m}_i\left(h_i + \frac{c_i^2}{2} + gz_i\right) - \dot{m}_c\left(h_c + \frac{c_c^2}{2} + gz_c\right).$$

Note how our use of enthalpy has simplified the rate of work term. In writing the control volume form of the equation we have assumed only one entering and one leaving stream, but this could be generalized to any number of inlet and exit streams.

In the special case of a steady-state flow,

$$\frac{d}{dt} = 0 \quad \text{and} \quad \dot{m}_i = \dot{m}_c = \dot{m}.$$

Applying this to Equation produces a form of the "Steady Flow Energy Equation" (SFEE),

$$\dot{Q}_{cv} - \dot{W}_{cv} + \dot{m}\left[\left(h_c + \frac{c_c^2}{2} + gz_c\right) - \left(h_i + \frac{c_i^2}{2} + gz_i\right)\right],$$

which has units of Joules per second. We could also divide by the mass flow to produce

$$q_{cv} - w_{cv} + \dot{m}\left(h_c + \frac{c_c^2}{2} + gz_c\right) - \left(h_i + \frac{c_i^2}{2} + gz_i\right),$$

which has units of Joules per second per kilogram. For problems of interest in aerospace applications the velocities are high and the term that is associated with changes in the elevation is small. From now on, we will neglect the *gz* terms unless explicitly stated.

STAGNATION TEMPERATURE AND STAGNATION ENTHALPY

Suppose that our steady flow control volume is a set of streamlines describing the flow up to the nose of a blunt object.

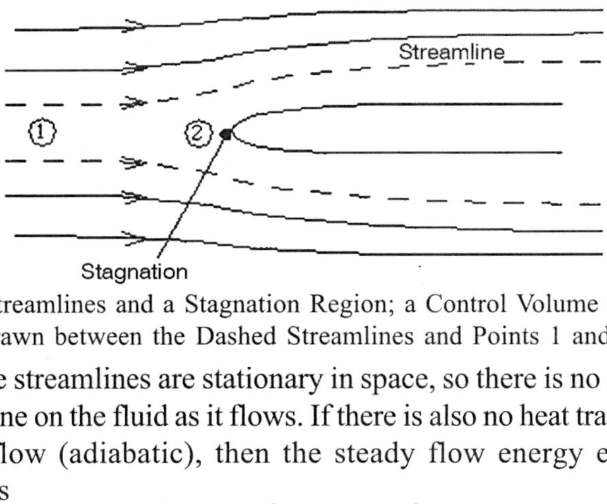

Fig. Streamlines and a Stagnation Region; a Control Volume can be Drawn between the Dashed Streamlines and Points 1 and 2

The streamlines are stationary in space, so there is no external work done on the fluid as it flows. If there is also no heat transferred to the flow (adiabatic), then the steady flow energy equation becomes

$$c_p T_2 + \frac{c_2^2}{2} = c_p T_1 + \frac{c_1^2}{1}.$$

The quantity that is conserved is defined as the stagnation temperature,

$$T_t \equiv T + \frac{c^2}{2c_p}$$

or

$$\frac{T_t}{T} = 1 + \frac{\gamma - 1}{2} M^2 \quad \text{using } a = \sqrt{\gamma RT}$$

where $M = c/a$ is the Mach number. The stagnation temperature is the temperature that the fluid would reach if it were brought to zero speed by a steady adiabatic process with no external work. Note that for any steady, adiabatic flow with no external work, the stagnation temperature is constant.

It is also convenient to define the stagnation enthalpy,

$$h_t = c_p T + \frac{c^2}{2}$$

which allows us to write the Steady Flow Energy Equation in a simpler form as

$$q_{1-2} - w_{s,1-2} = h_{t2} - h_{t1}.$$

Note that for a quasi-static adiabatic process

$$\frac{T_1}{T_2} = \left(\frac{p_1}{p_2}\right)^{\frac{\gamma-1}{\gamma}}$$

so we can write

$$\frac{T_1}{T} = \left(\frac{p_t}{p}\right)^{\frac{\gamma-1}{\gamma}}$$

and define the relationship between stagnation pressure and static pressure as

$$\frac{p_t}{p} \equiv \left(1 + \frac{\gamma - 1}{2} M^2\right)^{\frac{\gamma-1}{\gamma}},$$

where, the stagnation pressure is the pressure that the fluid would reach if it were brought to zero speed, via a steady, adiabatic, quasi-static process with no external work.

FRAME DEPENDENCE OF STAGNATION QUANTITIES

An area of common confusion is the frame dependence of stagnation quantities. The stagnation temperature and stagnation pressure are the conditions the fluid would reach if it were brought to zero speed relative to some reference frame, via a steady adiabatic process with no external work (for stagnation temperature) or a steady, adiabatic, reversible process with no external work (for stagnation pressure). Depending on the speed of the reference frame the stagnation quantities will take on different values.

Consider a high speed reentry vehicle traveling through the still atmosphere, which is at temperature, T. Let's place our reference frame on the vehicle and stagnate a fluid particle on the nose of the vehicle (carrying it along with the vehicle and thus essentially giving it kinetic energy). The stagnation temperature of the air in the vehicle frame is

$$T_t = T + \frac{c^2}{2c_p},$$

where c is the vehicle speed. The temperature the skin reaches (to first approximation) is the stagnation temperature and depends on the speed of the vehicle. Since re-entry vehicles travel fast, the skin temperature is much hotter than the atmospheric temperature. The atmospheric temperature, T, is not frame dependent, but the stagnation temperature, T_t, is.

The confusion comes about because T is usually referred to as the static temperature. In common language this has a similar meaning as "stagnation," but in fluid mechanics and thermodynamics static is used to label the thermodynamic properties of the gas (p, T, etc.), and these are not frame dependent.

Thus in our re-entry vehicle example, looking at the still atmosphere from the vehicle frame we see a stagnation temperature hotter than the atmospheric (static) temperature. If we look at the same still atmosphere from a stationary frame, the stagnation temperature is the same as the static temperature.

Example

For the case shown below, a jet engine is sitting motionless on the ground prior to take-off. Air is entrained into the engine by the compressor. The inlet can be assumed to be frictionless and adiabatic.

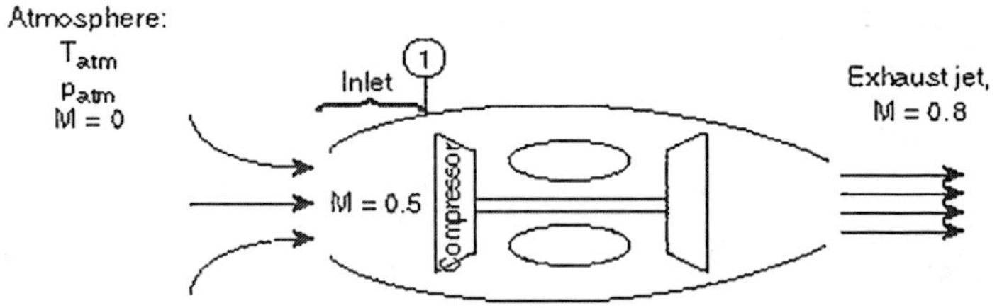

Fig. A Stationary Gas Turbine Drawing Air in from the Atmosphere

Considering the state of the gas within the inlet, prior to passage into the compressor, as state (1), and working in the reference frame of the motionless airplane:

- Is T_{t1} greater than, less than, or equal to T_{atm}?

The stagnation temperature of the atmosphere, $T_{t'atm}$, is equal to T_{atm} since it is moving the same speed as the reference frame (the motionless airplane). The steady flow energy equation tells us that if there is no heat or shaft work (the case for our adiabatic inlet) the stagnation enthalpy (and thus stagnation temperature for constant C_p) remains unchanged. Thus $T_{t1} = T_{t'atm} = T_{atm}$

- Is T_1 greater than, less than, or equal to T_{atm}?

If $T_{t1} = T_{t'atm}$ then $T_1 < T_{atm}$ since the flow is moving at station 1 and therefore some of the total energy is composed of kinetic energy (at the expense of internal energy, thus lowering T_1)

- Is p_{t1} greater than, less than, or equal to p_{atm}?
  Equal, by the same argument as 1.
- Is p_1 greater than, less than, or equal to p_{atm}?
  Less than, by the same argument as 2.

Steady Flow Energy Equation in Terms of Stagnation Enthalpy

The form of the "Steady Flow Energy Equation" (SFEE) that we will most commonly use is Equation written in terms of stagnation quantities, and neglecting chemical and potential energies,

Steady Flow Energy Equation: $\dot{Q}_{cv} - \dot{W}_{\text{shaft}} = \dot{m}\,(h_{t_c} - h_{ti})$. The steady flow energy equation finds much use in the analysis of power and propulsion devices and other fluid machinery. Note the prominent role of enthalpy.

EXAMPLE APPLICATIONS OF THE STEADY FLOW ENERGY EQUATION

Tank Filling

Using what we have just learned we can attack the tank filling problem solved from an alternate point of view using the control volume form of the first law. In this problem the shaft work is zero, and the heat transfer, kinetic energy changes, and potential energy changes are neglected. In addition there is no exit mass flow.

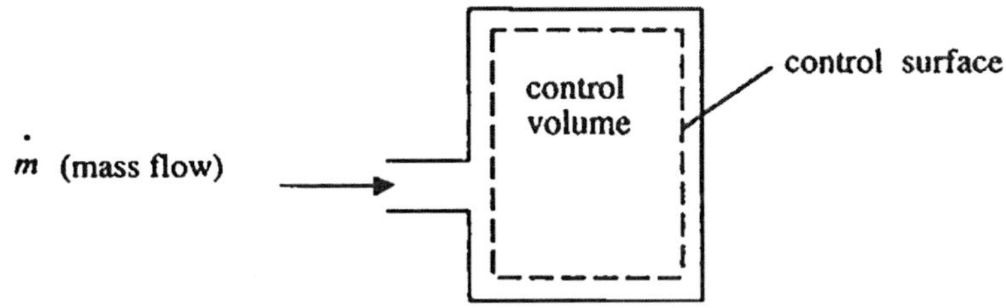

Fig. A Control Volume Approach to the Tank Filling Problem

The control volume form of the first law is therefore

$$\frac{dU}{dt} = \dot{m}_i h_i.$$

The equation of mass conservation is

$$\frac{dm}{dt} = \dot{m}_i.$$

Combining we have

$$\frac{dU}{dt} = \frac{dm}{dt} h_i.$$

Integrating from the initial time to the final time (the incoming enthalpy is constant) and using $U = mu$ gives the result $u_{final} = h_i = h_0$ as before.

Flow Through a Rocket Nozzle

A liquid bi-propellant rocket consists of a thrust chamber and nozzle and some means for forcing the liquid propellants into the chamber where they react, converting chemical energy to thermal energy.

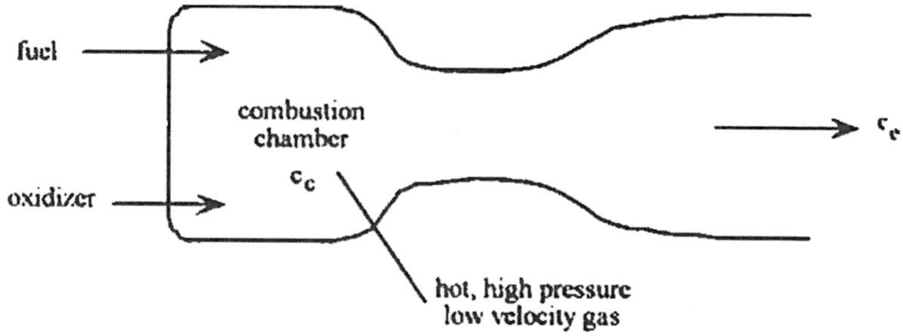

Fig. Flow Through a Rocket Nozzle

Once the rocket is operating we can assume that all of the flow processes are steady, so it is appropriate to use the steady flow energy equation. Also, for now we will assume that the gas behaves as a perfect gas with constant specific heats, though in general this is a poor approximation. There is no external work, and we assume that the flow is adiabatic. We define our control volume as going between location c, in the chamber, and location , at the exit, and then write the First Law as

$q_{c-c} - w_{s,\, c-c} = h_{tc} - h_{tc}$ which becomes $h_{tc} = h_{tc}$

or

$$c_p T_c + \frac{c_c^2}{2} = c_p T_c + \frac{c_c^2}{2},$$

Therefore

$$c_c = \sqrt{2c_p(T_c - T_c)}.$$

If we assume quasi-static, adiabatic expansion then

$$\frac{T_c}{T_c} = \left(\frac{p_c}{p_c}\right)^{\frac{\gamma-1}{\gamma}}$$

so

$$c_c = \sqrt{2c_p T_c\left[1 - \left(\frac{p_c}{p_c}\right)^{\frac{\gamma-1}{\gamma}}\right]}$$

T_c and p_c, the conditions in the combustion chamber, are set by propellants, and p_c is the external static pressure.

Power to Drive a Gas Turbine Compressor

The engine is designed to produce about 84,000 lbs of thrust at takeoff. The engine is a two-spool design. The fan and low pressure compressor are driven by the low pressure turbine. The high pressure compressor is driven by the high pressure turbine. We wish to find the total shaft work required to drive the compression system.

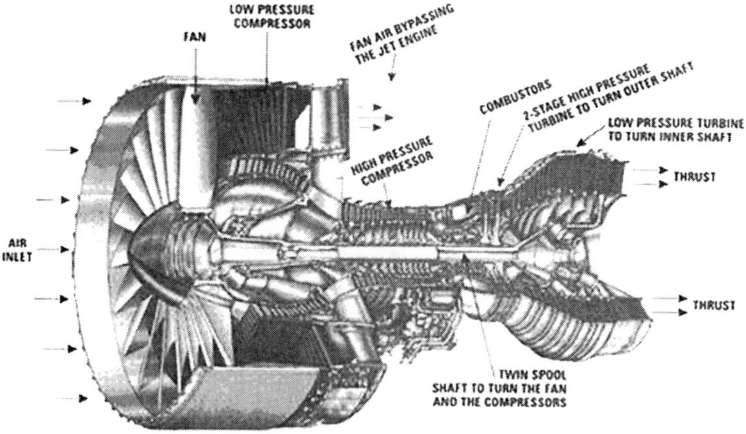

Fig. The Pratt and Whitney 4084 (drawing courtesy of Pratt and Whitney)

π_f = total pressure ratio across the fan ≈ 1.4

π_c = total pressure ratio across the fan + cinoressir ≈ 45

$\dot{m}_f = 610$ kg/s

$\dot{m}_{core} = 120$ kg/s

$T_{inlct} = 300$ K.

We define our control volume to encompass the compression system, from the front of the fan to the back of the fan and high pressure compressor, with the shaft cutting through the back side of the control volume. Heat transfer from the gas streams is negligible, so we write the First Law (steady flow energy equation) as:

$$\cancelto{0}{\dot{Q}} - \dot{W}_s = \dot{m}\,(h_{t2} - h_{t1}).$$

For this problem we must consider two streams, the fan stream, *f*, and the core stream, *c*:

$$-\dot{W}_s = \dot{m}_f \Delta h_{t,f} + \dot{m}_c \Delta h_{t,c}$$
$$= \dot{m}_f c_p \Delta T_{t,f} + \dot{m}_c c_p \Delta T_{t,c}$$

We obtain the temperature change by assuming that the compression process is quasi-static and adiabatic,

$$\frac{T_2}{T_1} = \left(\frac{p_2}{p_1}\right)^{\frac{\gamma-1}{\gamma}}$$

then

$$\left(\frac{T_{t2}}{T_{t1}}\right)_{\text{fan}} = \pi_f^{\frac{\gamma-1}{\gamma}} = 1.1 \Rightarrow \Delta T_{t,\,\text{fan}} = 30\,\text{K}$$

$$\left(\frac{T_{t2}}{T_{t1}}\right)_{\text{core}} = \pi_{\text{core}}^{\frac{\gamma-1}{\gamma}} = 3.0 \Rightarrow \Delta T_{t,\,\text{core}} = 600\,\text{K}$$

Substituting these values into the expression for the first law above, along with estimates of c_p, we obtain

$-\dot{W}_s$ = 610 kg/s × 30 K × 1008 J/kg-K + 120 kg/s × 600 K × 1008 J/kg-K

= – 91 × 106 J/s

= – 91 Megawatts negative sign implies work done on the fluid

Note that 1 Hp = 745 watts. If a car engine has ≈ 110 Hp = 8.2 × 10^4 watts, then the power needed to drive compressor is equivalent to 1,110 automobile engines. All of this power is generated by the low pressure and high pressure turbines.

APPLICATION OF FIRST LAW

SOME PROPERTIES OF ENGINEERING CYCLES; WORK AND EFFICIENCY

As preparation for our discussion of cycles (and as a foreshadowing of the second law), we examine two types of processes that concern interactions between heat and work. The first of these represents the conversion of work into heat. The second, which is much more useful, concerns the conversion of heat into work. The question we will pose is how efficient can this conversion be in the two cases.

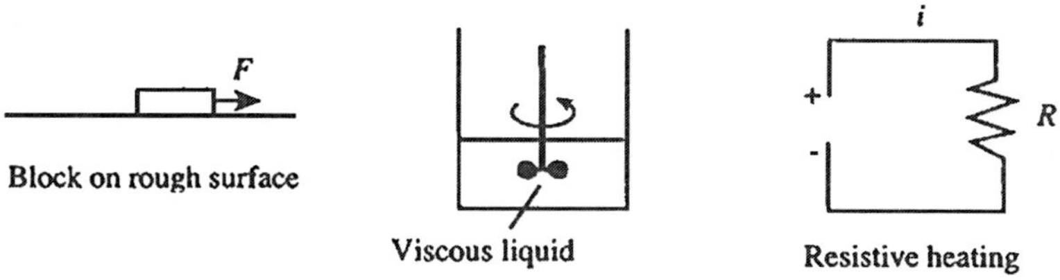

Fig. Examples of the Conversion of Work into Heat

The first is the pulling of a block on a rough horizontal surface by a force which moves through some distance. Friction resists the pulling. After the force has moved through the distance, it is removed. The block then has no kinetic energy and the same potential energy it had when the force started to act. If we measured the temperature of the block and the surface we would find that it was higher than when we started. (High temperatures can be reached if the velocities of pulling are high; this is the basis of inertia welding.) The work done to move the block has been converted totally to heat.

The second example concerns the stirring of a viscous liquid. There is work associated with the torque exerted on the shaft turning through an angle. When the stirring stops, the fluid comes to rest and there is (again) no change in kinetic or potential energy from the initial state. The fluid and the paddle wheels will be found to be hotter than when we started, however.

The final example is the passage of a current through a resistance. This is a case of electrical work being converted to heat, indeed it models operation of an electrical heater. All the examples in Figure have 100% conversion of work into heat. This 100% conversion could go on without limit as long as work were supplied. Is this true for the conversion of heat into work?

To answer the last question, we need to have some basis for judging whether work is done in a given process. One way to do this is to ask whether we can construct a way that the process could result in the raising of a weight in a gravitational field. If so, we can say "Work has been done." It may sometimes be difficult to make the link between a complicated thermodynamic process and the simple raising of a weight, but this is a rigorous test for the existence of work.

One example of a process in which heat is converted to work is the isothermal (constant temperature) expansion of an ideal gas. The system is the gas inside the chamber. As the gas expands, the piston does work on some external device. For an ideal gas, the internal energy is a function of temperature only, so that if the temperature is constant for some process the internal energy change is zero. To keep the temperature constant during the expansion, heat must be supplied. Because $\Delta U = 0$, the first law takes the form $Q = W$. This is a process that has 100% conversion of heat into work.

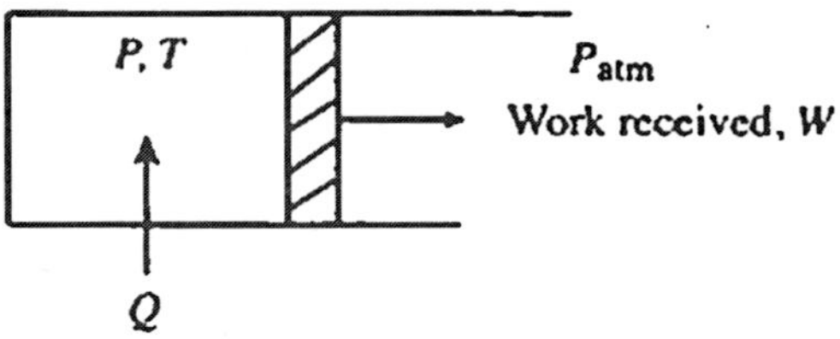

Fig. Isothermal Expansion

The work exerted by the system is given by

$$\int_1^2 PdV$$

where 1 and 2 denote the two states at the beginning and end of the process. The equation of state for an ideal gas is

$$P = NRT/V,$$

with N the number of moles of the gas contained in the chamber. Using the equation of state, the expression for work can be written as

Work during an isothermal expansion

$$= NRT \int_1^2 dV / V = NRT \ln\left(\frac{V_2}{V_1}\right).$$

For an isothermal process, PV = constant, so that $P_1/P_2 = V_2/V_1$. The work can be written in terms of the pressures at the beginning and end as

$$\text{Work during an isothermal expansion} = NRT \ln\left(\frac{V_2}{V_1}\right).$$

The lowest pressure to which we can expand and still receive work from the system is atmospheric pressure. Below this, we would have to do work on the system to pull the piston out further. There is thus a bound on the amount of work that can be obtained in the isothermal expansion; we cannot continue indefinitely. For a power or propulsion system, however, we would like a source of continuous power, in other words a device that would give power or propulsion as long as fuel was added to it. To do this, we need a series of processes where the system does not progress through a one-way transition from an initial state to a different final state, but rather cycles back to the initial state. What is looked for is in fact a thermodynamic cycle for the system.

We define several quantities for a cycle:

- *QA* is the heat absorbed by the system.
- *QR* is the heat rejected by the system.
- *W*i s the net work done by the system.

The cycle returns to its initial state, so the overall energy change, ΔU, is zero. The net work done by the system is related to the magnitudes of the heat absorbed and the heat rejected by

$$W = \text{Net work} = Q_A - Q_R.$$

The thermal efficiency of the cycle is the ratio of the work done to the heat absorbed. (Efficiencies are often usefully portrayed as "What you get" versus "What you pay for." Here what we get is work and what we pay for is heat, or rather the fuel that generates the heat.) In terms of the heat absorbed and rejected, the thermal efficiency is

$$\eta = \text{thermal efficiency} = \frac{\text{Work done}}{\text{Heat absorbed}} = \frac{Q_A - Q_R}{Q_A} = 1 - \frac{Q_R}{Q_A}.$$

The thermal efficiency can only be 100% (complete conversion of heat into work) if $Q_R = 0$; a basic question is what is the maximum thermal efficiency for any arbitrary cycle? We examine this for several cases, including the Carnot cycle and the Brayton (or Joule) cycle, which is a model for the power cycle in a jet engine.

GENERALIZED REPRESENTATION OF THERMODYNAMIC CYCLES

Before we examine individual heat engines, note that all heat engines can be represented generally as a transfer of heat from a high temperature reservoir to a device, which does work on the surroundings, followed by a rejection of heat from that device to a low temperature reservoir.

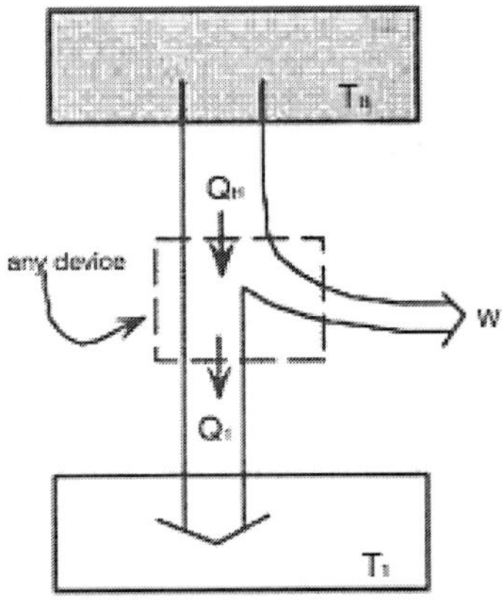

Fig. A generalized heat engine

THE CARNOT CYCLE

It has four processes. There are two adiabatic reversible legs and two isothermal reversible legs. We can construct a Carnot cycle with many different systems, but the concepts can be shown using a familiar working fluid, the ideal gas. The system can be regarded as a chamber enclosed by a piston and filled with this ideal gas.

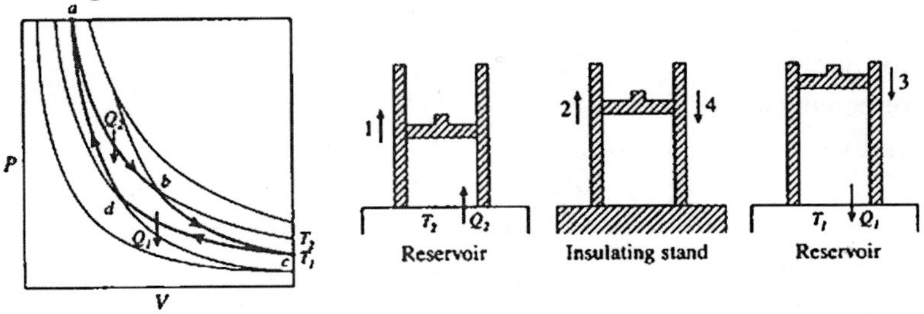

Fig. Carnot Cycle Thermodynamic Diagram on Left and Schematic of the Different Stages in the Cycle for a System Composed of an Ideal Gas on the Right

The four processes in the Carnot cycle are:

- The system is at temperature T_2 at state *a*. It is brought in contact with a heat reservoir, which is just a liquid or solid mass of large enough extent such that its temperature does not change appreciably when some amount of heat is transferred to the system. In other words, the heat reservoir is a constant temperature source (or receiver) of heat. The system then undergoes an isothermal expansion from *a* to *b*, with heat absorbed Q_2.

- At state *b*, the system is thermally insulated (removed from contact with the heat reservoir) and then let expand to *c*. During this expansion the temperature decreases to T_1. The heat exchanged during this part of the cycle, Q_{bc} = 0)
- At state the system is brought in contact with a heat reservoir at temperature T_1. It is then compressed to state *d*, rejecting heat Q_1 in the process.
- Finally, the system is compressed adiabatically back to the initial state *a*. The heat exchange $Q_{da} = 0$.

The thermal efficiency of the cycle is given by the definition

$$\eta = 1 - \frac{Q_R}{Q_A} = 1 + \frac{Q_1}{Q_2}.$$

In this equation, there is a sign convention implied. The quantities Q_A, Q_R as defined are the magnitudes of the heat absorbed and rejected. The quantities Q_1, Q_2 on the other hand are defined with reference to heat received by the system. In this example, the former is negative and the latter is positive. The heat absorbed and rejected by the system takes place during isothermal processes and we already know what their values are:

$$Q_2 = W_{ab} = NRT_2\,[\ln\,(V_b\,/\,V_a)],$$

$$Q_1 = W_{cb} = NRT_1\,[\ln\,(V_d\,/\,V_c)] = -\,[\ln\,(V_c\,/\,V_d)]. \text{ } (Q_1 \text{ is negative.})$$

The efficiency can now be written in terms of the volumes at the different states as

$$\eta = 1 + \frac{T_1\,[\ln\,(V_d\,/\,V_c)]}{T_2[\ln\,(V_b\,/\,V_a)]}.$$

The path from states *b* to *c* and from *a* to *d* are both adiabatic and reversible. For a reversible adiabatic process we know that PV^{γ} = constant. Using the ideal gas equation of state, we have $TV^{\gamma-1}$ = constant. Along curve b-c , therefore, $T_2 V_b^{\gamma-1} = T_1 V_c^{\gamma-1}$. Along the curve d-a, $T_2 V_a^{\gamma-1} = T_1 V_d^{\gamma-1}$. Thus,

$$\left(\frac{V_d}{V_c}\right)^{\gamma-1} = \frac{(T_2/T_1)}{(T_2/T_1)}\left(\frac{V_a}{V_b}\right)^{\gamma-1}, \text{ which means that } \frac{V_d}{V_c} = \frac{V_a}{V_b}.$$

Comparing the expression for thermal efficiency equations show two consequences. First, the heats received and rejected are related to the temperatures of the isothermal parts of the cycle by

$$\frac{Q_1}{T_1} + \frac{Q_2}{T_2} = 0.$$

Second, the efficiency of a Carnot cycle is given compactly by

$$\eta = 1 - \frac{T_1}{T_2}. \text{ Carnot cycle efficiency.}$$

The efficiency can be 100% only if the temperature at which the heat is rejected is zero. The heat and work transfers to and from the system.

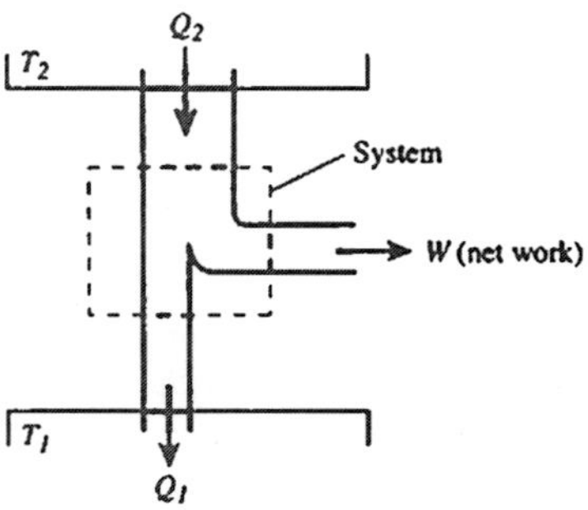

Fig. Work and Heat Transfers in a Carnot Cycle between Two Heat Reservoirs

REFRIGERATORS AND HEAT PUMPS

The Carnot cycle has been used for power, but we can also run it in reverse. If so, there is now net work into the system and net heat out of the system. There will be a quantity of heat Q_2 rejected at the higher temperature and a quantity of heat Q_1 absorbed at the lower temperature. The former of these is negative according to our convention and the latter is positive. The result is that work is done on the system, heat is extracted from a low temperature source and rejected to a high temperature source.

The words "low" and "high" are relative and the low temperature source might be a crowded classroom on a hot day, with the heat extraction being used to cool the room. In this mode of operation the cycle works as a refrigerator or heat pump. "What we pay for" is the work, and "what we get" is the amount of heat extracted. A metric for devices of this type is the coefficient of performance, defined as

$$\text{Coefficient of performance} = \frac{Q1}{-W} = \frac{Q_1}{-(Q_1 + Q_2)}.$$

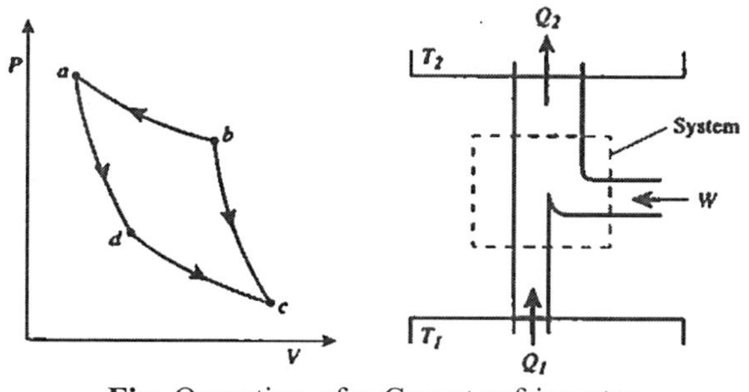

Fig. Operation of a Carnot refrigerator

For a Carnot cycle we know the ratios of heat in to heat out when the cycle is run forward and, since the cycle is reversible, these ratios are the same when the cycle is run in reverse. The coefficient of performance is thus given in terms of the absolute temperatures as

$$\text{Coefficiency of performance} = \frac{T_1}{T_2 - T_1}.$$

This can be much larger than unity.

The Carnot cycles that have been drawn are based on ideal gas behaviour. For different working media, however, they will look different. We will see an example when two-phase situations.

What is the same whatever the medium is the efficiency for all Carnot cycles operating between the same two temperatures.

REFRIGERATOR HARDWARE

Typically the thermodynamic system in a refrigerator analysis will be a working fluid, a refrigerant, that circulates around a loop. The internal energy (and temperature) of the refrigerant is alternately raised and lowered by the devices in the loop. The working fluid is colder than the refrigerator air at one point and hotter than the surroundings at another point. Thus heat will flow in the appropriate direction, as shown by the two arrows in the heat exchangers.

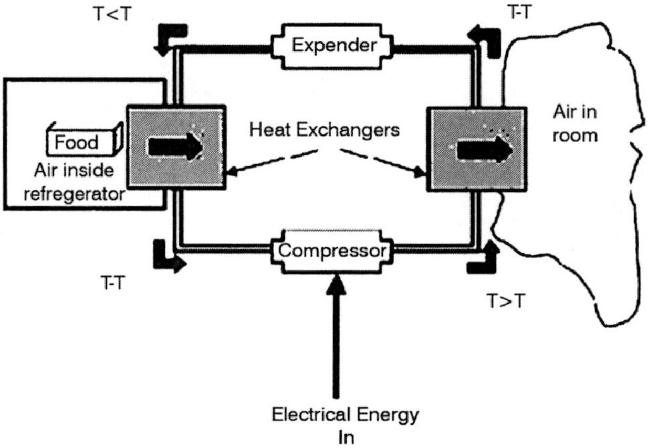

Fig. Schematic of a Domestic Refrigerator

Starting in the upper right hand corner of the diagram, we describe the process in more detail. First the refrigerant passes through a small turbine or through an expansion valve. In these devices, work is done by the refrigerant so its internal energy is lowered to a point where the temperature of the refrigerant is lower than that of the air in the refrigerator.

A heat exchanger is used to transfer energy from the inside of the refrigerator to the cold refrigerant. This lowers the internal energy of the inside and raises the internal energy of the refrigerant. Then a pump or compressor is used to do work on the refrigerant, adding additional energy to it and thus further raising its internal energy. Electrical energy is used to drive the pump or compressor. The internal energy of the refrigerant is raised to a point where its temperature is hotter than the temperature of the surroundings.

The refrigerant is then passed through a heat exchanger (often coils at the back of the refrigerator) so that energy is transferred from the refrigerant to the surroundings.

As a result, the internal energy of the refrigerant is reduced and the internal energy of the surroundings is increased. It is at this point where the internal energy of the contents of the refrigerator and the energy used to drive the compressor or pump are transferred to the surroundings.

The refrigerant then continues on to the turbine or expansion valve, repeating the cycle.

THE INTERNAL COMBUSTION ENGINE (OTTO CYCLE)

The Otto cycle is a set of processes used by spark ignition internal combustion engines (2-stroke or 4-stroke cycles). These engines a) ingest a mixture of fuel and air, b) compress it, c) cause it to react, thus effectively adding heat through converting chemical energy into thermal energy, d) expand the combustion products, and then e) eject the combustion products and replace them with a new charge of fuel and air.

Intake stroke, gasoline vapour and air drawn into engine ($5 \rightarrow 1$).

- Compression stroke, p, $T_{increase}$ ($1 \rightarrow 2$).
- Combustion (spark), short time, essentially constant volume ($2 \rightarrow 3$). Model: heat absorbed from a series of reservoirs at temperatures T_2 to T_3.
- Power stroke: expansion ($3 \rightarrow 4$).
- Valve exhaust: valve opens, gas escapes.
- ($4 \rightarrow 1$) Model: rejection of heat to series of reservoirs at temperatures T_4 to T_1.
- Exhaust stroke, piston pushes remaining combustion products out of chamber ($1 \rightarrow 5$).

We model the processes as all acting on a fixed mass of air contained in a piston-cylinder arrangement.

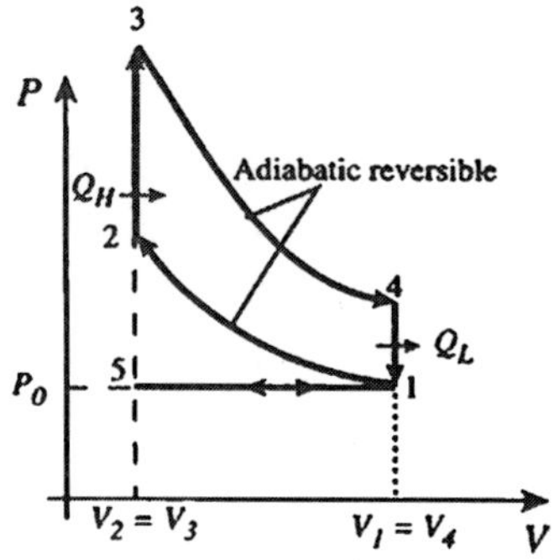

Fig. The ideal Otto Cycle

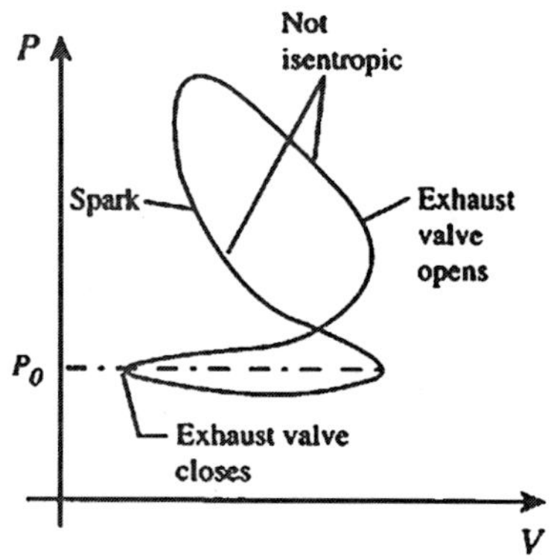

Fig. Sketch of an Actual Otto Cycle

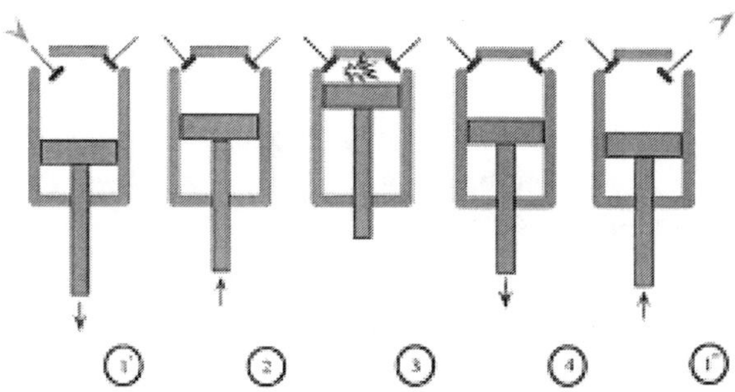

Fig. Piston and Valves in a Four-stroke Internal Combustion Engine

EFFICIENCY OF AN IDEAL OTTO CYCLE

The starting point is the general expression for the thermal efficiency of a cycle:

$$\eta = \frac{\text{work}}{\text{heat input}} = \frac{Q_H + Q_L}{Q_H} = 1 + \frac{Q_L}{Q_H}.$$

The convention, as previously, is that heat exchange is positive if heat is flowing into the system or engine, so Q_L is negative. The heat absorbed occurs during combustion when the spark occurs, roughly at constant volume. The heat absorbed can be related to the temperature change from state 2 to state 3 as:

$$\begin{aligned} Q_H &= Q_{23} = \Delta U_{23} \quad (W_{23} = 0) \\ &= \int_{T_2}^{T_3} C_v dT = C_v(T_3 - T_2). \end{aligned}$$

The heat rejected is given by (for a perfect gas with constant specific heats)

$$Q_L = Q_{41} = \Delta U_{41} = C_v(T_1 - T_4).$$

Substituting the expressions for the heat absorbed and rejected in the expression for thermal efficiency yields

$$\eta = 1 - \frac{T_4 - T_1}{T_3 - T_2}.$$

We can simplify the above expression using the fact that the processes from 1 to 2 and from 3 to 4 are isentropic:

$$T_4 V_1^{\gamma-1} = T_3 V_2^{\gamma-1}, \quad T_1 V_1^{\gamma-1} = T_2 V_2^{\gamma-1}$$

$$(T_4 - T_1)\, V_1^{\gamma-1} = (T_3 - T_2) V_2^{\gamma-1}$$

$$\frac{T_4 - T_1}{T_3 - T_2} = \left(\frac{V_2}{V_1}\right)^{\gamma-1}.$$

The quantity $V_1/V_2 = r$ is called the compression ratio. In terms of compression ratio, the efficiency of an ideal Otto cycle is:

$$\eta_{\text{otto}} = 1 - \frac{1}{(V_1 / V_2)^{\gamma-1}} = 1 - \frac{1}{r^{\gamma-1}}.$$

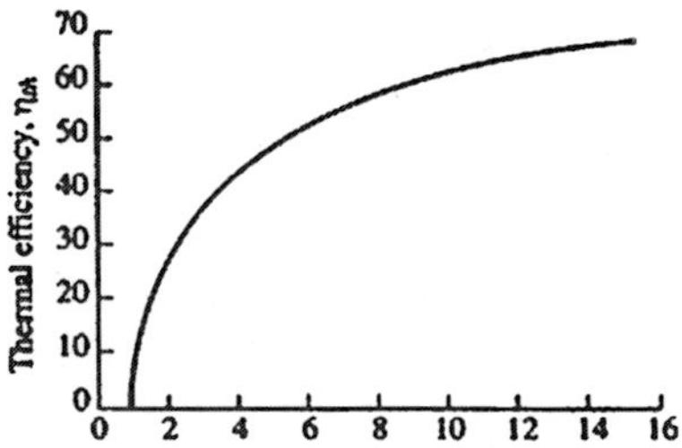

Fig. Ideal Otto Cycle Thermal Efficiency

The ideal Otto cycle efficiency is shown as a function of the compression ratio. As the compression ratio, r, increases, η_{Otto} increases, but so does T_2. If T_2 is too high, the mixture will ignite without a spark (at the wrong location in the cycle).

ENGINE WORK, RATE OF WORK PER UNIT ENTHALPY FLUX

The non-dimensional ratio of work done (the power) to the enthalpy flux through the engine is given by

$$\frac{\text{Power}}{\text{Enthalpy flux}} = \frac{\dot{W}}{\dot{m}c_p T_1} = \frac{\dot{Q}_{23} h_{\text{Otto}}}{\dot{m}c_p T_1}.$$

There is often a desire to increase this quantity, because it means a smaller engine for the same power. The heat input is given by

$$\dot{Q}_{23} = \dot{m}_{\text{fucl}} \Delta h_{\text{fucl}},$$

where

- Δh_{fucl} is the heat of reaction, i.e. the chemical energy liberated per unit mass of fuel,
- $\dot{m}_{\text{fuel}}$ is the fuel mass flow rate.

The non-dimensional power is

$$\frac{\dot{W}}{\dot{m}_{cp} T_1} = \frac{\dot{m}_{\text{fucl}}}{\dot{m}} \frac{\Delta h_{\text{fucl}}}{c_p T_1} \left[1 - \frac{1}{r^{\gamma-1}} \right].$$

The quantities in this equation, evaluated at stoichiometric conditions are:

$$\frac{\dot{m}_{\text{fucl}}}{\dot{m}} \approx \frac{1}{15}$$

$$\frac{\Delta h_{\text{fuel}}}{c_p T_1} \approx \frac{4\times10^7}{10^3\times288},$$

so

$$\frac{\dot{W}}{\dot{m}c_p T_1} \approx 9\left[1-\frac{1}{r^{\gamma-1}}\right].$$

DIESEL CYCLE

The Diesel cycle is a compression ignition (rather than spark ignition) engine. Fuel is sprayed into the cylinder at P_2 (high pressure) when the compression is complete, and there is ignition without a spark.

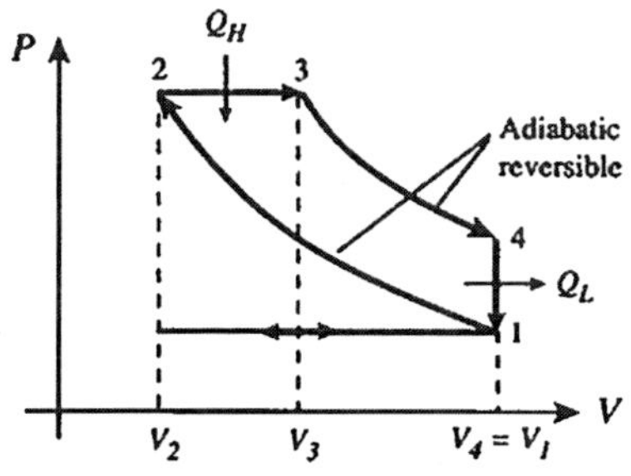

Fig. The ideal Diesel cycle

The thermal efficiency is given by:

$$\eta_{\text{Diesel}} = 1+\frac{Q_L}{Q_H} = 1+\frac{C_v(T_1-T_4)}{C_p(T_3-T_2)}$$

$$= 1-\frac{T_1(T_4/T_1-1)}{\gamma T_2(T_3/T_2-1)}.$$

This cycle can operate with a higher compression ratio than the Otto cycle because only air is compressed and there is no risk of auto-ignition of the fuel.

Although for a given compression ratio the Otto cycle has higher efficiency, because the Diesel engine can be operated to higher compression ratio, the engine can actually have higher efficiency than an Otto cycle when both are operated at compression ratios that might be achieved in practice.

BRAYTON CYCLE

The Brayton cycle (or Joule cycle) represents the operation of a gas turbine engine. The cycle consists of four processes, alongside a sketch of an engine:

- *a* – *b* Adiabatic, quasi-static (or reversible) compression in the inlet and compressor;
- *b* – *c* Constant pressure fuel combustion (idealized as constant pressure heat addition);
- *c* – *d* Adiabatic, quasi-static (or reversible) expansion in the turbine and exhaust nozzle, with which we
- Take some work out of the air and use it to drive the compressor, and
- Take the remaining work out and use it to accelerate fluid for jet propulsion, or to turn a generator for electrical power generation;
- *d* – *a* Cool the air at constant pressure back to its initial condition.

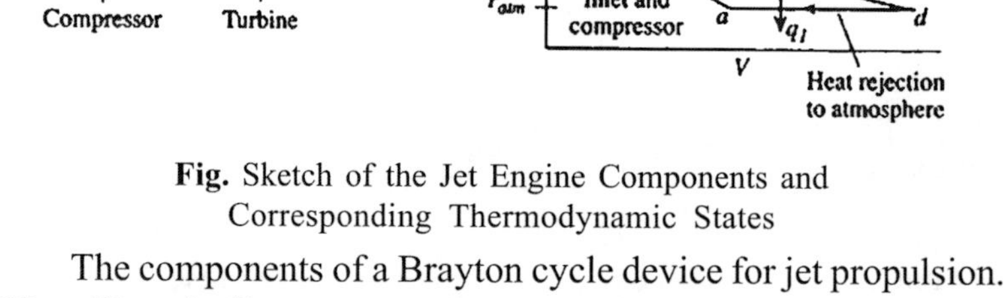

Fig. Sketch of the Jet Engine Components and Corresponding Thermodynamic States

The components of a Brayton cycle device for jet propulsion. We will typically represent these components schematically. In practice, real Brayton cycles take one of two forms. "open" cycle, where the working fluid enters and then exits the device.

This is the way a jet propulsion cycle works. The alternative, a closed cycle, which recirculates the working fluid. Closed cycles are used, in space power generation.

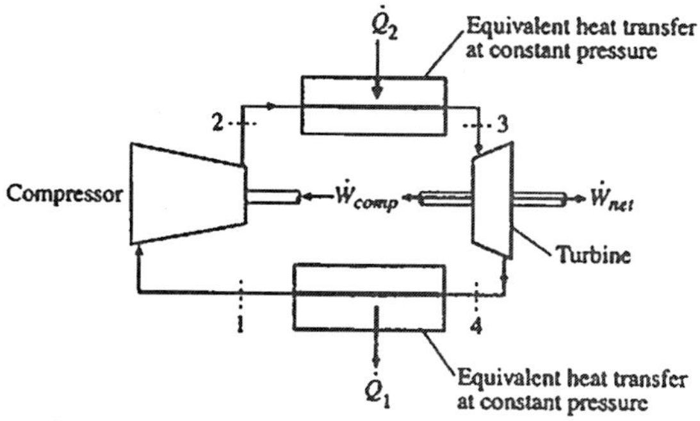

Fig. Thermodynamic model of gas turbine engine cycle for power generation

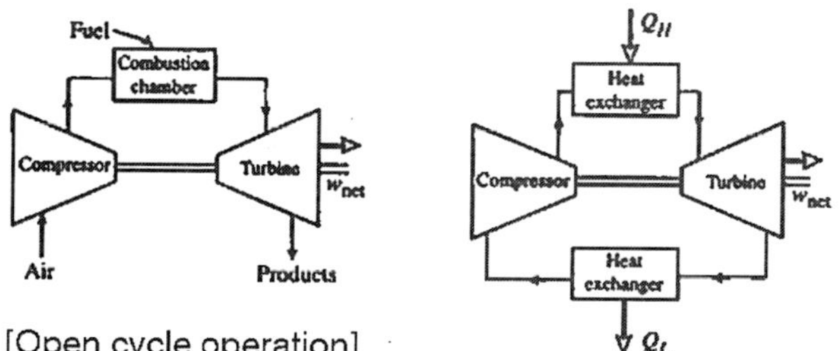

Fig. Options for operating Brayton cycle gas turbine engines

WORK AND EFFICIENCY

The objective now is to find the work done, the heat absorbed, and the thermal efficiency of the cycle. Tracing the path shown around the cycle from *a-b-c-d* and back to *a*, the first law gives (writing the equation in terms of a unit mass),

$$\Delta u_{a-b-c-d-a} = 0 = q_2 + q_1 - w.$$

Here Δu is zero because is a function of state, and any cycle returns the system to its starting state. The net work done is therefore $w = q_2 + q_1$, where q_1, $q_2$are defined as heat received by the system (q_1 is negative). We thus need to evaluate the heat transferred in processes *b-c* and *d-a*.

For a constant pressure, quasi-static process the heat exchange per unit mass is

$$dh = c_p dT = dq, \text{ or } [dq]_{\text{constant}}\ p = dh.$$

We can see this by writing the first law in terms of enthalpy or by remembering the definition of c_p. The heat exchange can be expressed in terms of enthalpy differences between the relevant states. Treating the working fluid as a perfect gas with constant specific heats, for the heat addition from the combustor,

$$q_2 = h_c - h_b = c_p\ (T_c - T_b).$$

The heat rejected is, similarly,

$$q_1 = h_a - h_d = c_p\ (T_a - T_d).$$

The net work per unit mass is given by

$$\text{Net work per unit mass} = q_1 + q_2 = c_p[\ (T_c - T_b) + (T_a - T_d)].$$

The thermal efficiency of the Brayton cycle can now be expressed in terms of the temperatures:

$$\eta = \frac{\text{Net work}}{\text{Heat in}} = \frac{c_p[(T_c - T_b) - (T_d - T_a)}{c_p(T_c - T_b)}$$

$$= 1 - \frac{(T_a - T_a)}{(T_c - T_b)} = 1 - \frac{T_a(T_d / T_a - 1)}{T_b(T_c / T_b - 1)}.$$

To proceed further, we need to examine the relationships between the different temperatures. We know that points a and d are on a constant pressure process as are points b and c, and $P_a = P_d$; $P_b = P_c$. The other two legs of the cycle are adiabatic and reversible, so

$$\frac{P_d}{P_c} = \frac{P_a}{P_b} \Rightarrow \left(\frac{T_d}{T_c}\right)^{\gamma/(\gamma-1)} = \left(\frac{T_a}{T_b}\right)^{\gamma/(\gamma-1)}.$$

Therefore $T_d/T_c = T_d/T_b$, or, finally, $T_d/T_a = T_c/T_b$. Using this relation in the expression for thermal efficiency, Equation yields an expression for the thermal efficiency of a Brayton cycle:

$$\text{Ideal Bratyton cycle efficiency: } \eta_B = 1 - \frac{T_a}{T_b} = 1 - \frac{T_{\text{atmospheric}}}{T_{\text{compressor exit}}}.$$

The temperature ratio across the compressor, $T_b/T_a = TR$. In terms of compressor temperature ratio, and using the relation for an adiabatic reversible process we can write the efficiency in terms of the compressor (and cycle) pressure ratio, which is the parameter commonly used:

$$\eta_B = 1 - \frac{1}{TR} = 1 - \frac{1}{PR^{(\gamma-1)/\gamma}}.$$

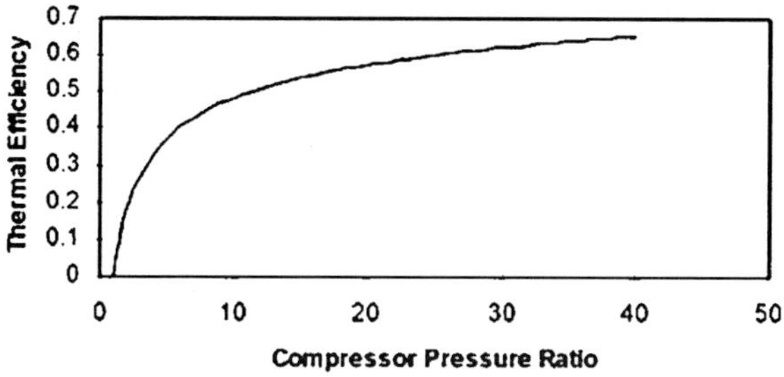

Fig. Trend of Brayton Cycle Thermal Efficiency with Compressor Pressure Ratio

Equation says that for a high cycle efficiency, the pressure ratio of the cycle should be increased. The history of aircraft engine pressure ratio versus entry into service, and it can be seen that there has been a large increase in cycle pressure ratio. The thermodynamic concepts apply to the behaviour of real aerospace devices!

GAS TURBINE TECHNOLOGY AND THERMODYNAMICS

The turbine entry temperature, T_c, is fixed by materials technology and cost. (If the temperature is too high, the blades fail.) The relation between the gas temperature coming into the turbine blades and the blade melting temperature.

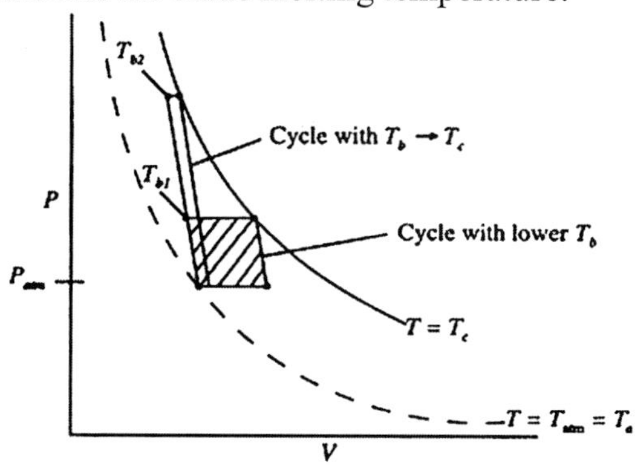

Fig. Efficiency and Work of Two Brayton Cycle Engines

The problem is posed which shows two Brayton cycles. For maximum efficiency we would like *TR* as high as possible. This means that the compressor exit temperature approaches the turbine entry temperature. The net work will be less than the heat received; as $T_b \to T_c$ the heat received approaches zero and so does the net work.

The net work in the cycle can also be expressed as $\int P dv$, evaluated in traversing the cycle. This is the area enclosed by the curves, which is seen to approach zero as $T_b \to T_c$.

The conclusion from either of these arguments is that a cycle designed for maximum thermal efficiency is not very useful in that the work (power) we get out of it is zero.

A more useful criterion is that of maximum work per unit mass (maximum power per unit mass flow). This leads to compact propulsion devices. The work per unit mass is given by:

$$\text{Work/unit mass} = c_p[(T_c - T_b) - (T_d - T_a)],$$

where T_c is the maximum turbine inlet temperature (a design constraint) and T_a is atmospheric temperature. The design variable is the compressor exit temperature, T_b, and to find the maximum as this is varied, we differentiate the expression for work with respect to:

$$\frac{d\,\text{Work}}{dT_b} = c_p\left[\frac{dT_c}{dT_b} - 1 - \frac{dT_d}{dT_b} + \frac{dT_a}{dT_b}\right].$$

The first and the fourth terms on the right hand side of the above equation are both zero (the turbine entry temperature is fixed, as is the atmospheric temperature). The maximum work occurs where the derivative of work with respect to T_b is zero:

$$\frac{d\,\text{Work}}{dT_b} = 0 = -1 - \frac{dT_d}{dT_b}.$$

To use Equation, we need to relate T_d and T_b. We know that

$$\frac{T_d}{T_a} = \frac{T_c}{T_b} \quad \text{or} \quad T_d = \frac{T_a T_c}{T_b}.$$

Hence,

$$\frac{dT_d}{dT_b} = -\frac{T_a T_c}{T_b^2}.$$

Plugging this expression for the derivative into Equation gives the compressor exit temperature for maximum work as $T_b = \sqrt{T_a T_c}$. In terms of temperature ratio,

$$\text{Compressor temperature ratio for maximum work: } \frac{T_b}{T_a} = \sqrt{\frac{T_c}{T_a}}.$$

The condition for maximum work in a Brayton cycle is different than that for maximum efficiency. The role of the temperature ratio can be seen if we examine the work per unit mass which is delivered at this condition:

$$\text{Work/unit mass} = c_p\left[T_c - \sqrt{T_a T_c} - \frac{T_a T_c}{\sqrt{T_a T_c}} + T_a\right].$$

Ratioing all temperatures to the engine inlet temperature,

$$\text{Work/unit mass} = c_p T_a\left[\frac{T_c}{T_a} - 2\sqrt{\frac{T_c}{T_a}} + 1\right].$$

To find the power the engine can produce, we need to multiply the work per unit mass by the mass flow rate:

$$\text{Power} = \dot{m} c_p T_a\left[\frac{T_c}{T_a} - 2\sqrt{\frac{T_c}{T_a}} + 1\right];$$ Maximum power for an ideal Brayton cycle.

(The units are $\frac{kg}{s}\frac{J}{kg-K}K = \frac{J}{s} =$ Watts.)

The trend of work output vs. compressor pressure ratio, for different temperature ratios TR = T_c/T_a.

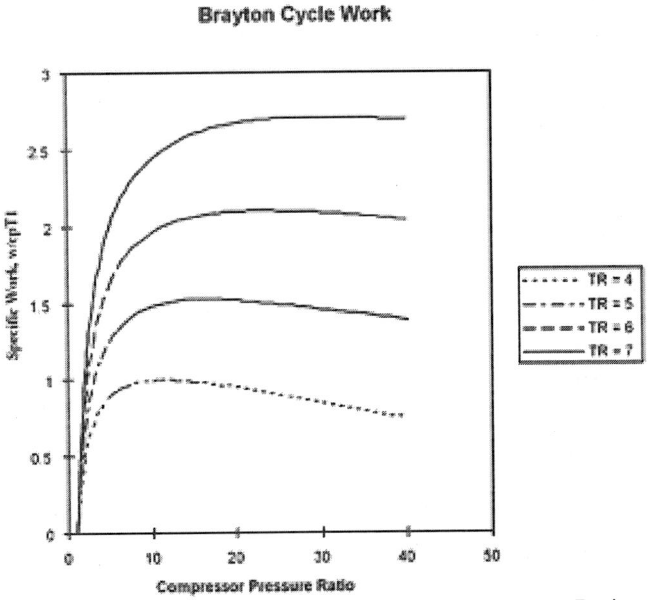

Fig. Trend of Cycle work with Compressor Pressure Ratio, for Different Temperature Ratios

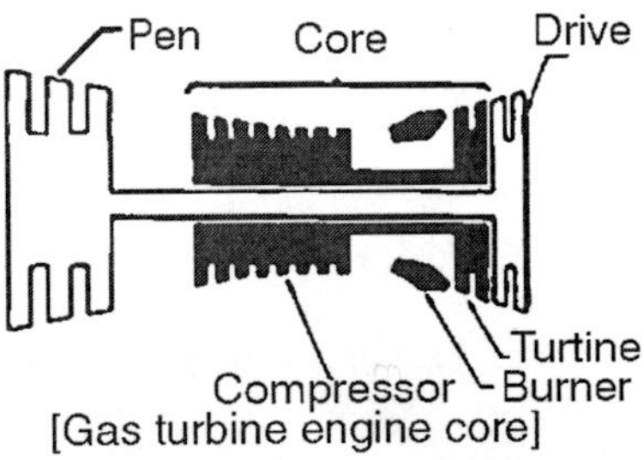

Fig. Aeroengine Core Power

The expression for power of an ideal cycle compared with data from actual jet engines. The gas turbine engine layout including the core (compressor, burner, and turbine). Thecore power for a number of different engines as a function of the turbine rotor entry temperature. The equation in the figure for horsepower (HP) is the same as that which we just derived, except for the conversion factors. The analysis not only shows the qualitative trend very well but captures much of the quantitative behaviour too. A final comment on Brayton cycles concerns the value of the thermal efficiency. The Brayton cycle thermal efficiency contains the ratio of the compressor exit temperature to atmospheric temperature, so that the ratio is not based on the highest temperature in the cycle, as the Carnot efficiency is. For a given maximum cycle temperature, the Brayton cycle is therefore less efficient than a Carnot cycle.

Brayton Cycle for Jet Propulsion: The Ideal Ramjet

A schematic of a ramjet is given in Figure.

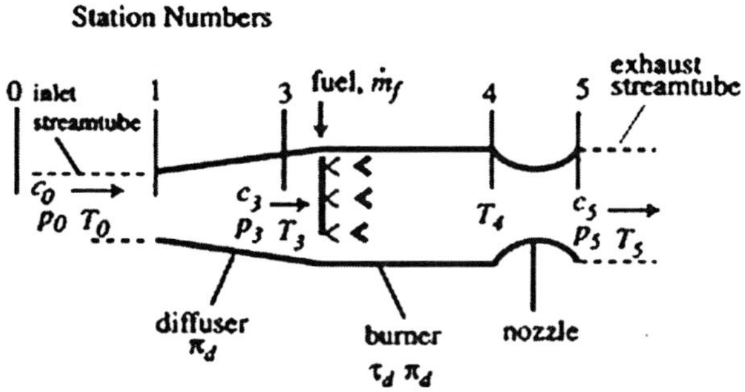

Fig. Ideal Ramjet

In the ramjet there are "no moving parts." The processes that occur in this propulsion device are:

- $0 \rightarrow 3$: Isentropic diffusion and compression, with a decrease in Mach number, $M_0 \rightarrow M_3 << 1$.
- $3 \rightarrow 4$: Constant pressure combustion.
- $4 \rightarrow 5$: Isentropic expansion through the nozzle.

The ramjet thermodynamic cycle efficiency can be written in terms of flight Mach number, M_0, as follows:

$$\eta_{\text{Brayton}} = 1 - \frac{T_0}{T_{\text{compressor exit}}} = 1 - \frac{T_0}{T_3} = 1 - \frac{T_0}{T_{t0}}$$

and
$$\frac{T_0}{T_{t0}} = \frac{1}{1 + \frac{\gamma - 1}{2} M_0^2},$$

so
$$\eta_{\text{Brayton}} = \frac{\frac{\gamma - 1}{2} M_0^2}{1 + \frac{\gamma - 1}{2} M_0^2}.$$

REFERENCES

- A Guide to Physics: Thermodynamics, Statistical Physics, and Quantum Mechanics by Gerald D. Mahan, Boris E. Nadgorny, and Max Dresden.
- Mixing and Excess Thermodynamic Properties (Physical sciences data) by Jaime Wisniak and Abraham Tamir.
- Statistical mechanics, R.P. Feynman, W. A. Benjamin.
- The Language of Science by Sidney B. Cahn.
- Statistical Mechanics, K. Huang,

8

The Second Law of Thermodynamics

DIFFERENCE BETWEEN FREE EXPANSION OF A GAS AND REVERSIBLE ISOTHERMAL EXPANSION

The difference between reversible and irreversible processes is brought out through examination of the isothermal expansion of an ideal gas. The question to be asked is what is the difference between the "free expansion" of a gas and the isothermal expansion against a piston? To answer this, we address the steps that we would have to take to reverse, in other words, to undo the process. By free expansion, we mean the unrestrained expansion of a gas into a volume. Initially all the gas is in the volume designated as V_1 with the rest of the insulated enclosure a vacuum. The total volume (V_1 plus the evacuated volume) is V_2. At a given time a hole is opened in the partition and the gas rushes through to fill the rest of the enclosure.

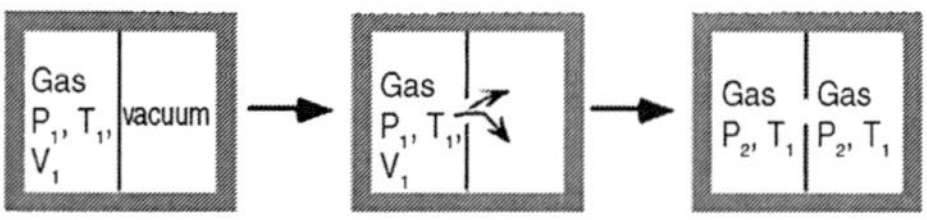

Fig. Free Expansion

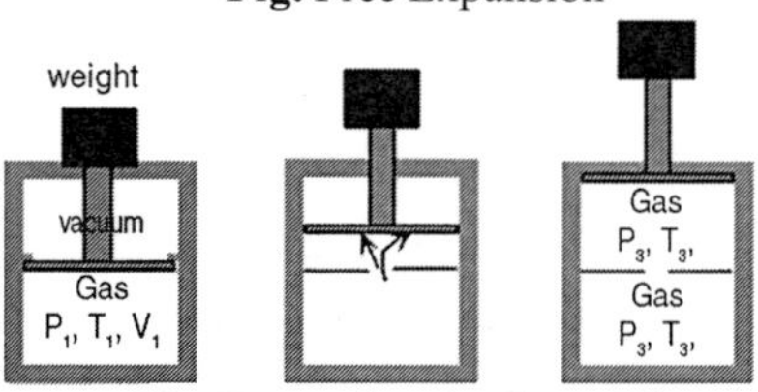

Fig. Expansion Against a Piston

During the expansion there is no work exchanged with the surroundings because there is no motion of the boundaries. The enclosure is insulated so there is no heat exchange.

The first law tells us therefore that the internal energy is constant ($\Delta U = 0$). For an ideal gas, the internal energy is a function of temperature only so that the temperature of the gas before the free expansion and after the expansion has been completed is the same. Characterizing the before and after states:

- Before: State 1, $V = V_1$, $T = T_1$
- After: State 2, $V = V_2$, $T = T_1$

$Q = W = 0$, so there is no change in the surroundings.

To restore the original state, i.e., to go back to the original volume at the same temperature ($V_2 \rightarrow V_1$ at constant $T = T_1$) we can compress the gas isothermally (using work from an external agency). We can do this in a quasi-equilibrium manner, with $P_{\text{system}} \approx P_{\text{external}}$. If so the work that we need to do is W = $\int_1^2 PdV$. We have evaluated the work in a reversible isothermal expansion, and we can apply the arguments to the case of a reversible isothermal compression. The work done on the system to go from state "2" to state "1" is

$$W = \text{Work done on system} = NRT_1 \ln\left(\frac{V_2}{V_1}\right).$$

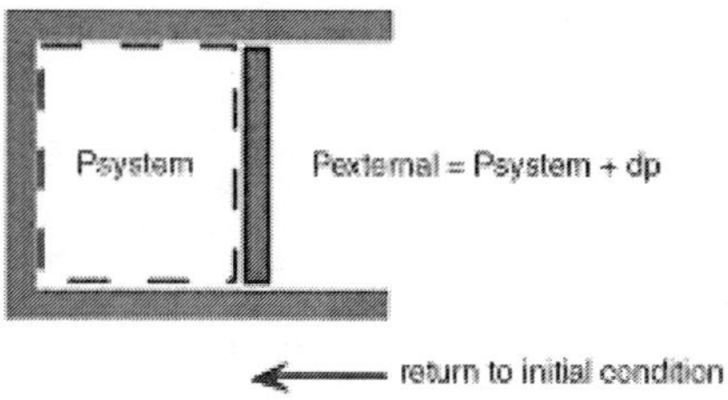

Fig. Returning the Free Expansion to its Initial Condition

From the first law, this amount of heat must also be rejected from the gas to the surroundings if the temperature of the gas is to remain constant. A schematic of the compression process, in terms of heat and work exchanged.

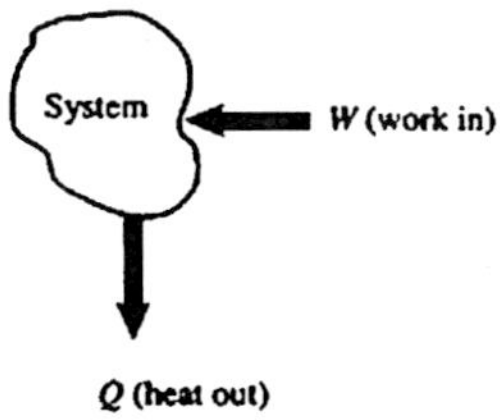

Fig. Work and Heat Exchange in the Reversible Isothermal Compression Process

At the end of the combined process (free expansion plus reversible compression):

- The system has been returned to its initial state (no change in system state).
- The surroundings (us!) did work on the system of magnitude *W*.
- The surroundings received an amount of heat, *Q*, which is equal to *W*.
- The sum of all of these events is that we have converted an amount of work, *W*, into an amount of heat, *Q*, with *W* and *Q* numerically equal in Joules.

The net effect is the same as if we let a weight fall and pull a block along a rough surface. There is 100% conversion of work into heat.

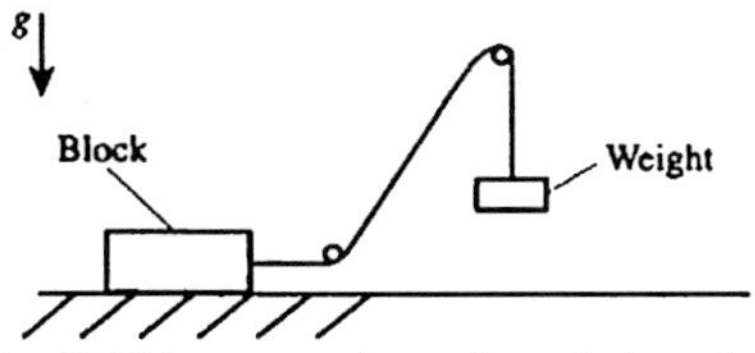

Fig. 100% conversion of work into heat

The results of the free expansion can be contrasted against a process of isothermal expansion against a pressure which is slightly different than that of the system.

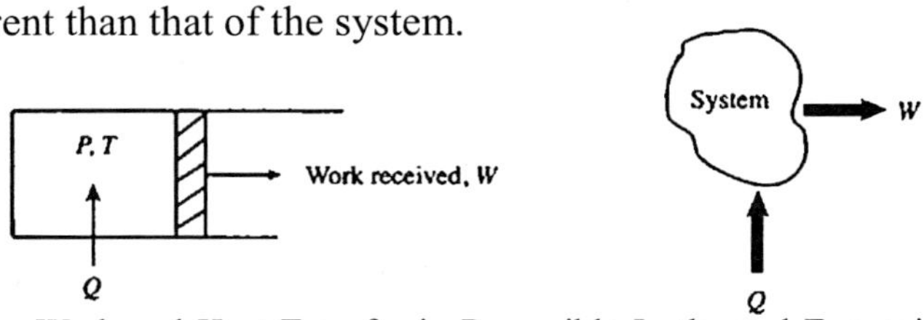

Fig. Work and Heat Transfer in Reversible Isothermal Expansion

During the expansion, work is done on the surroundings of magnitude W = $\int PdV$, where P can be taken as the system pressure. As evaluated in Equation, the magnitude of the work done by the system is $W = NRT_1 \ln (V_2/V_1)$. At the end of the isothermal expansion, therefore:

- The surroundings have received work W.
- The surroundings have given up heat, Q, numerically equal to W.

We now wish to restore the system to its initial state, just as we did in the free expansion. To do this we need to do work on the system and extract heat from the system, just as in the free expansion. In fact, because we are doing a transition between the same states along the same path, the work and heat exchange are the same as those for the compression process examined just above.

The overall result when we have restored the system to the initial state, however, is quite different for the reversible expansion than for the free expansion. For the reversible expansion, the work we need to do on the system to compress it has the same magnitude as the work we received during the expansion process. Indeed, we could raise a weight during the expansion and then allow it to be lowered during the compression process. Similarly the heat put into the system by us (the surroundings) during the expansion process has the same magnitude as the heat received by us during the compression process. The result is that when the system has been restored back to its initial state, so have the surroundings. There is no trace of the overall process on either the system or the surroundings. That is another meaning of the word "reversible."

FEATURES OF REVERSIBLE PROCESSES

Reversible processes are idealizations or models of real processes. One familiar and widely used example is Bernoulli's equation, which you saw in Unified. They are extremely useful for defining limits to system or device behaviour, for enabling identification of areas in which inefficiencies occur, and in giving targets for design. An important feature of a reversible process is that, depending on the process, it represents the maximum work that can be extracted in going from one state to another, or the minimum work that is needed to create the state change.

Let us consider processes that do work, so that we can show that the reversible one produces the maximum work of all possible processes between two states. Suppose we have a thermally insulated cylinder that holds an ideal gas. The gas is contained by a thermally insulated massless piston with a stack of many small weights on top of it. Initially the system is in mechanical and thermal equilibrium.

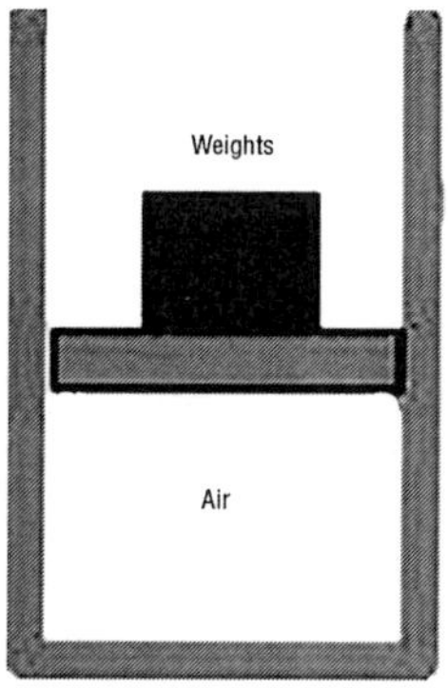

Fig. A piston with weights on top

Let us consider the following three processes:

- All of the weights are removed from the piston instantaneously and the gas expands until its volume is increased by a factor of four (a free expansion).
- Half of the weight is removed from the piston instantaneously, the system is allowed to double in volume, and then the remaining half of the weight is instantaneously removed from the piston and the gas is allowed to expand until its volume is again doubled.
- Each small weight is removed from the piston one at a time, so that the pressure inside the cylinder is always in equilibrium with the weight on top of the piston. When the last weight is removed, the volume has increased by a factor of four.

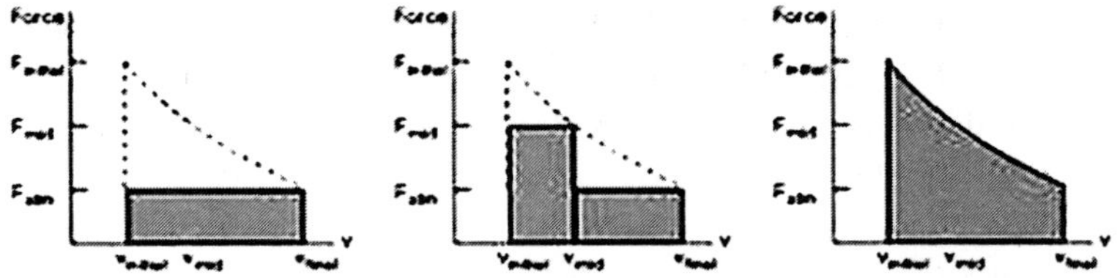

Fig. Getting the Most Work Out of a System Requires that the Work be Extracted Reversibly

- Maximum work (proportional to the area under these curves) is obtained for the quasi-static expansion.

To reiterate:

- The work done by a system during a reversible process is the maximum work we can get.
- The work done on a system in a reversible process is the minimum work we need to do to achieve that state change.

A process must be quasi-static (quasi-equilibrium) to be reversible. This means that the following effects must be absent or negligible:

- Friction: If $P_{\text{external}} \neq P_{\text{system}}$ we would have to do net work to bring the system from one volume to another and return it to the initial condition
- Free (unrestrained) expansion.
- Heat transfer through a finite temperature difference.

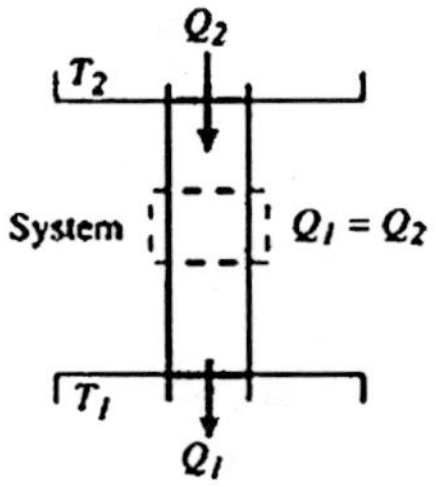

Fig. Heat Transfer Across a Finite Temperature Difference

Suppose we have heat transfer from a high temperature to a lower temperature. How do we restore the situation to the initial conditions? One thought would be to run a Carnot refrigerator to get an amount of heat, Q, from the lower temperature reservoir to the higher temperature reservoir. We could do this but the surroundings, again us, would need to provide some amount of work (which we could find using our analysis of the Carnot refrigerator). The net (and only) result at the end of the combined process would be a conversion of an amount of work into heat.

For reversible heat transfer from a heat reservoir to a system, the temperatures of the system and the reservoir must be $T_{\text{heat reservoir}} = T_{\text{system}} \pm dT$. In other words the difference between the temperatures of the two entities involved in the heat transfer process can only differ by an infinitesimal amount, dT.

While all natural processes are irreversible to some extent, it cannot be emphasized too strongly that there are a number of engineering situations where the effect of irreversibility can be neglected and the reversible process furnishes an excellent approximation to reality.

The second law, which is the next topic we address, allows us to make a quantitative statement concerning the irreversibility of a given physical process.

THE SECOND LAW OF THERMODYNAMICS

STATEMENTS OF THE LAWS OF THERMODYNAMICS

As a further aid in familiarization with the second law of thermodynamics and the idea of entropy, we draw an analogy with statements made previously concerning quantities that are closer to experience. In particular, we wish to present once more the Zeroth and First Laws of thermodynamics and use the same framework for the Second Law. In this so-called "axiomatic formulation," the parallels between the Zeroth, First and Second Laws will be made explicit.

Zeroth Law

Zeroth Law: There exists for every thermodynamic system in equilibrium a property called temperature. Equality of temperature is a necessary and sufficient condition for thermal equilibrium. The Zeroth law thus defines a property (temperature) and describes its behaviour.

First Law

Observations also show that for any system there is a property called the energy. The First Law asserts that one must associate such a property with every system.

First Law: There exists for every thermodynamic system a property called the energy. The change of energy of a system is equal to the mechanical work done on the system in an adiabatic process. In a non-adiabatic process, the change in energy is equal to the heat added to the system minus the mechanical work done by the system.

On the basis of experimental results, therefore, one is led to assert the existence of two new properties, the temperature and internal energy, which do not arise in ordinary mechanics. In a similar way, a further remarkable relationship between heat and temperature will be established, and a new property, the entropy, defined. Although this is a much less familiar property, it is to be stressed that the general approach is quite like that used to establish the Zeroth and First Laws.

A general principle and a property associated with any system are extracted from experimental results. Viewed in this way, the entropy should appear no more mystical than the internal energy. The increase of entropy in a naturally occurring process is no less real than the conservation of energy.

Second Law

Although all natural processes must take place in accordance with the First Law, the principle of conservation of energy is, by itself, inadequate for an unambiguous description of the behaviour of a system. Specifically, there is no mention of the familiar observation that every natural process has in some sense a preferred direction of action. The flow of heat occurs naturally from hotter to colder bodies, in the absence of other influences, but the reverse flow certainly is not in violation of the First Law. So far as that law is concerned, the initial and final states are symmetrical in a very important respect.

The Second Law is essentially different from the First Law;

the two principles are independent and cannot in any sense be deduced from one another. Thus, the concept of energy is not sufficient, and a new property must appear. This property can be developed, and the Second Law introduced, in much the same way as the Zeroth and First Laws were presented. By examination of certain observational results, one attempts to extract from experience a law which is supposed to be general; it is elevated to the position of a fundamental axiom to be proved or disproved by subsequent experiments. Within the structure of classical thermodynamics, there is no proof more fundamental than observations. A statement which can be adopted as the Second Law of thermodynamics is:

Second Law: There exists for every thermodynamic system in equilibrium an extensive scalar property called the entropy, S, such that in an infinitesimal reversible change of state of the system, $dS = dQ/T$, where T is the absolute temperature and dQ is the amount of heat received by the system. The entropy of a thermally insulated system cannot decrease and is constant if and only if all processes are reversible.

As with the Zeroth and First Laws, the existence of a new property is asserted and its behaviour is described.

REVERSIBLE PROCESSES

In the course of this development, the idea of a completely reversible process is central, and we can recall the definition, "a process is called completely reversible if, after the process has occurred, both the system and its surroundings can be wholly restored by any means to their respective initial states". Especially, it is to be noted that the definition does not, in this form, specify that the reverse path must be identical with the forward path. If the initial states can be restored by any means whatever, the process is by definition completely reversible.

If the paths are identical, then one usually calls the process (of the system) reversible, or one may say that the state of the system follows a reversible path. In this path between two equilibrium states:

- The system passes through the path followed by the equilibrium states only.
- The system will take the reversed path by a simple reversal of the work done and heat added.

Reversible processes are idealizations not actually encountered. However, they are clearly useful idealizations. For a process to be completely reversible, it is necessary that it be quasi-static and that there be no dissipative influences such as friction and diffusion. The precise (necessary and sufficient) condition to be satisfied if a process is to be reversible is the second part of the Second Law.

The criterion as to whether a process is completely reversible must be based on the initial and final states. In the form presented above, the Second Law furnishes a relation between the properties defining the two states, and thereby shows whether a natural process connecting the states is possible.

COMBINED FIRST AND SECOND LAW EXPRESSIONS

The first law, written in a form that is always true:

$$dU = dQ - dW.$$

For reversible processes only, work or heat may be rewritten as

$$dW = PdV,$$

$$dQ = TdS,$$

Substitution leads to other forms of the first law true for reversible processes only:

$$dU = dQ - \underline{PdV}, \text{ substituted for a reversible } dQ.$$

$$dQ = \underline{TdS} - dW, \text{ substituted for a reversible } dQ.$$

(If the substance has other work modes, e.g., stress, strain,

$$dU = dQ - PdV - XdY,$$

where X is a pressure-like quantity, and Y is a volume-like quantity.)

Substituting for both dW and dQ in terms of state variables,

$$dU = TdS - PdV \quad \text{Always true.}$$

The above is always true because it is a relation between properties and is now independent of process.

In terms of specific quantities:

$dU = TdS - PdV$ Combined first and second law (a) or Gibbs equation (a).

The combined first and second law expressions are often more usefully written in terms of the enthalpy, or specific enthalpy, h = u + Pv,

$$dh = \underline{du} + Pdv + vdP$$

$$= \underline{Tds - Pdv} + Pdv + vdP, \text{ using in the first law}.$$

$$dh = Tds + vdP.$$

Or, since v = 1/ρ,

$$dh = Tds + \frac{dP}{\rho}$$ Combined first and second law (b) or Gibbs equation (b).

In terms of enthalpy (rather than specific enthalpy) the relation is

$$dH = TdS + VdP.$$

ENTROPY CHANGES IN AN IDEAL GAS

Many aerospace applications involve flow of gases (e.g., air) and we thus examine the entropy relations for ideal gas behaviour. The starting point is form (a) of the combined first and second law,

$$du = Tds - Pdv.$$

For an ideal gas, $du = c_v dT$. Thus

$$Tds = c_v dT + Pdv \quad \text{or} \quad ds = c_v \frac{dT}{T} + \frac{P}{T} dv,$$

Using the equation of state for an ideal gas ($Pv = RT$), we can write the entropy change as an expression with only exact differentials:

$$ds = c_v \frac{dT}{T} + R\frac{dv}{v}.$$

We can think of Equation as relating the fractional change in temperature to the fractional change of volume, with scale factors

c_v and R; if the volume increases without a proportionate decrease in temperature (as in the case of an adiabatic free expansion), then increases. Integrating Equation between two states "1" and "2":

$$\Delta s = s_2 - s_1 = \int_{T_1}^{T_2} c_v \frac{dT}{T} + R \int_{v_1}^{v_2} \frac{dv}{v}.$$

For a perfect gas with constant specific heats

$$\Delta s = s_2 - s_1 = c_v \ln\left(\frac{T_2}{T_1}\right) + R \ln\left(\frac{v_2}{v_1}\right).$$

In non-dimensional form (using $R/c_v = (\gamma - 1)$)

$$\frac{\Delta s}{c_v} = \ln\left(\frac{T_2}{T_1}\right) + (\gamma - 1)\ln\left(\frac{v_2}{v_1}\right),$$ Entropy change of a perfect gas.

Equation is in terms of specific quantities. For N moles of gas,

$$\frac{\Delta s}{c_v} = N\left[\ln\left(\frac{T_2}{T_1}\right) + (\gamma - 1)\ln\left(\frac{V_2}{V_1}\right)\right].$$

This expression gives entropy change in terms of temperature and volume. We can develop an alternative form in terms of pressure and volume, which allows us to examine an assumption we have used. The ideal gas equation of state can be written as

$$\ln P + \ln v = \ln R + \ln T.$$

Taking differentials of both sides yields

$$\frac{dP}{P} + \frac{dv}{v} = \frac{dT}{T}.$$

Using the above equation in Equation, and making use of the relations $c_p = c_v + R$; $c_p/c_v = \gamma$, we find

$$ds = cv\left[\frac{dP}{P} + \frac{dv}{v}\right] + R\frac{dv}{v},$$

or

$$\frac{ds}{c_v} = \frac{dP}{P} + \gamma\frac{dv}{v}.$$

Integrating between two states 1 and 2

$$\frac{\Delta s}{c_v} = \ln\left(\frac{P_2}{P_1}\right) + \gamma \ln\left(\frac{v_2}{v_1}\right) = \ln\left[\frac{P_2}{P_1}\left(\frac{v_2}{v_1}\right)^{\gamma}\right].$$

Using both sides of Equation as exponents we obtain

$$\frac{P_2 v_2^{\gamma}}{P_1 v_1^{\gamma}} = [Pv^{\gamma}]_1^2 = e^{\Delta s / c_v}.$$

Equation describes a general process. For the specific situation in which $\Delta s = 0$, i.e., the entropy is constant, we recover the expression $Pv^{\gamma} =$ constant. It was stated that this expression applied to a reversible, adiabatic process. Through use of the second law, a deeper meaning to the expression, and to the concept of a reversible adiabatic process, in that both are characteristics of a constant entropy, or isentropic, process.

CALCULATION OF ENTROPY CHANGE IN SOME BASIC PROCESSES

- Heat transfer from, or to, a heat reservoir.

A heat reservoir is a constant temperature heat source or sink. Because the temperature is uniform, there is no heat transfer across a finite temperature difference and the heat exchange is reversible. From the definition of entropy ($dS = dQ_{rev}/T$),

$$\Delta S = \frac{Q}{T},$$

where Q is the heat into the reservoir (defined here as positive if heat flows into the reservoir.)

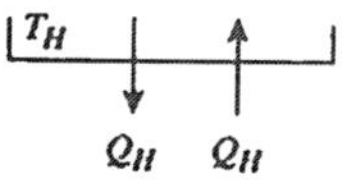

Fig. Heat Transfer From/to a Heat Reservoir

- Heat transfer between two heat reservoirs

The entropy change of the two reservoirs in Figure is the sum of the entropy change of each. If the high temperature reservoir is at T_H and the low temperature reservoir is at T_L, the total entropy change is

$$\Delta S = \left(\frac{-Q}{T_H}\right) + \left(\frac{Q}{T_L}\right) = \frac{Q}{T_H T_L}(T_H - T_L).$$

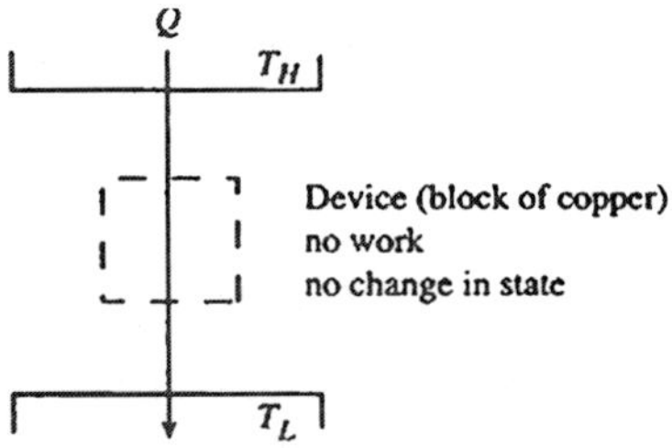

Fig. Heat Transfer between Two Reservoirs

The second law says that the entropy change must be equal to or greater than zero. This corresponds to the statement that heat must flow from the higher temperature source to the lower temperature source.

- Possibility of obtaining work from a single heat reservoir

We can regard the process proposed in Figure as the absorption of heat, Q, by a device or system, operating in a cycle, rejecting no heat, and producing work. The total entropy change is the sum of the change in the reservoir, the system or device, and the surroundings. The entropy change of the reservoir is $\Delta S = -Q/T_H$. The entropy change of the device is zero, because we are considering a complete cycle (return to initial state) and entropy is a function of state. The surroundings receive work only so the entropy change of the surroundings is zero. The total entropy change is

$$\Delta S_{\text{total}} = \Delta S_{\text{reservoir}} + \Delta S_{\text{device}} + \Delta S_{\text{surroundings}}$$
$$= -Q/T_H + = 0 + 0$$

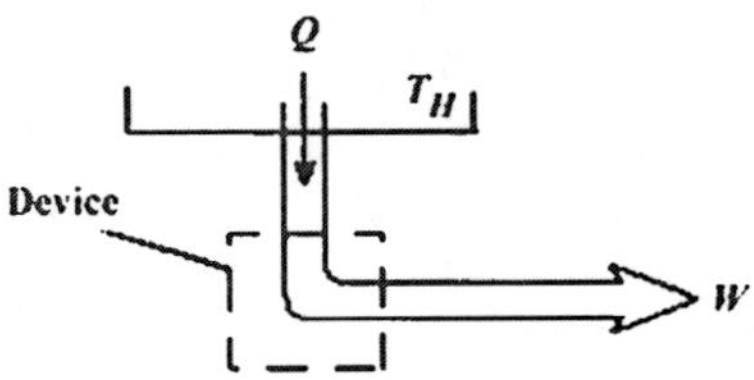

Fig. Work from a single heat reservoir

The total entropy change in the proposed process is thus less than zero,

$$\Delta S_{\text{total}} < 0,$$

which is not possible. The second law thus tells us that we cannot get work from a single reservoir only. The "only" is important; it means without any other changes occurring.

- Entropy changes in the "hot brick problem"

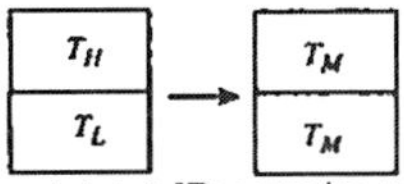

[Temperature equalization of two bricks] [Reservoirs used in reversible state transformation]

Fig. The "Hot Brick" Problem

We can examine in a more quantitative manner the changes that occurred when we put the two bricks together, as depicted. The process by which the two bricks come to the same temperature is not a reversible one, so we need to devise a reversible path. To do this imagine a large number of heat reservoirs at varying temperatures spanning the range $T_H - dT, \ldots, T_L + dT$. The bricks are put in contact with them sequentially to raise the temperature of one and lower the temperature of the other in a reversible manner. The heat exchange at any of these steps is $dQ = CdT$. For the high temperature brick, the entropy change is:

$$\Delta S_{\text{hot brick}} = \int_{T_H}^{T_M} \frac{CdT}{T} = C \ln\left(\frac{T_M}{T_H}\right),$$

where C is the heat capacity of the brick (J/kg). This quantity is less than zero. For the cold brick,

$$\Delta S_{\text{cold brick}} = \int_{T_L}^{T_M} \frac{CdT}{T} = C \ln\left(\frac{T_M}{T_L}\right).$$

The entropy change of the two bricks is

$$\Delta S_{\text{brick}} = C\left[\ln\left(\frac{T_M}{T_H}\right) + \ln\left(\frac{T_M}{T_L}\right)\right] = C \ln\left(\frac{T_M^2}{T_H T_L}\right) > 0.$$

The process is not reversible.

- Difference between the free expansion and the reversible isothermal expansion of an ideal gas

The essential difference between the free expansion in an insulated enclosure and the reversible isothermal expansion of an ideal gas can also be captured clearly in terms of entropy changes. For a state change from initial volume and temperature $V_1 T_1$, to final volume and (the same) temperature $V_2 T_1$, the entropy change is

$$\Delta S = \int_1^2 dS = \int_1^2 \frac{dU}{T} + \int_1^2 \frac{PdV}{T},$$

or, making use of the equation of state and the fact that $dU = 0$ for an isothermal process,

$$\Delta S = NR \ln\left(\frac{V_2}{V_1}\right).$$

This is the entropy change that occurs for the free expansion as well as for the isothermal reversible expansion processes- entropy changes are state changes and the two system final and end states are the same for both processes.

For the free expansion:

$$\Delta S_{\text{system}} = NR \ln\left(\frac{V_2}{V_1}\right); \quad \Delta S_{\text{surrundings} = 0.}$$

There is no change in the entropy of the surroundings because there is no interaction between the system and the surroundings. The total entropy change is therefore,

$$\Delta S_{\text{total}} = \Delta S_{\text{system}} + \Delta S_{\text{surroundings}} = NR \ln\left(\frac{V_2}{V_1}\right) > 0.$$

There are several points to note from this result:

- $\Delta S_{\text{total}} > 0$ so the process is not reversible.
- $\Delta S_{\text{system}} > \int_1^2 dQ/T = 0$; the equality between ΔS and dQ/T is only for a reversible process.
- There is a direct connection between the work needed to restore the system to the original state and the entropy change:

$$W = NRT \ln\left(\frac{V_2}{V_1}\right) = T\Delta S_{2-1}.$$

The $T\Delta S$ quantity has a physical meaning as "lost work" in the sense of work which we lost the opportunity to utilize.

For the reversible isothermal expansion:

The entropy is a state variable so the entropy change of the system is the same as before. In this case, however, heat is transferred to the system from the surroundings ($Q_{\text{surroundings}} < 0$) so that $\Delta S_{\text{surroundings}} = Q_{\text{surroundings}}/T < 0$.

The heat transferred from the surroundings, however, is equal to the heat received by the system: $Q_{\text{surroundings}} = Q_{\text{system}} = W$.

$$\Delta S_{\text{surroundings}} = \frac{Q_{\text{surroundings}}}{T} = -\frac{W}{T} = -NR\ln\left(\frac{V_2}{V_1}\right).$$

The total change in entropy (system plus surroundings) is therefore

$$\Delta S_{\text{total}} = \Delta S_{\text{system}} + \Delta S_{\text{surroundings}} = \frac{Q}{T} - \frac{Q}{T} = 0.$$

The reversible process has zero total change in entropy.

APPLICATIONS OF THE SECOND LAW

THE THERMODYNAMIC TEMPERATURE SCALE

The considerations of Carnot cycles in this section have not mentioned the working medium. They are thus not limited to an ideal gas and hold for Carnot cycles with any medium. Earlier we derived the Carnot efficiency with an ideal gas as a medium and the temperature definition used in the ideal gas equation was not essential to the thermodynamic arguments. More specifically, we can define a thermodynamic temperature scale that is independent of the working medium. To see this, which has three reversible cycles. There is a high temperature heat reservoir at T_1 and a low temperature heat reservoir at T_3. For any two temperatures T_1, T_2, the ratio of the magnitudes of the heat absorbed and rejected in a Carnot cycle has the same value for all systems.

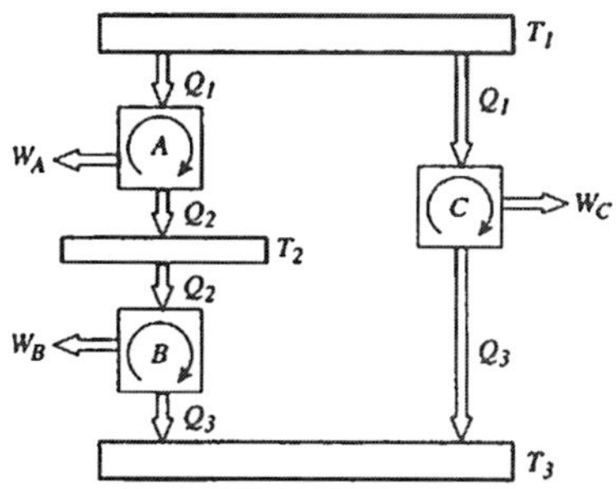

Fig. Arrangement of Heat Engines to Demonstrate the Thermodynamic Temperature Scale

We choose the cycles so Q_1 is the same for A and C. Also Q_3 is the same for B and C. For a Carnot cycle

$$\eta = 1 + \frac{Q_L}{Q_H} = F(T_L, T_H); \eta \text{ is only a function of temperature.}$$

Also

$$\frac{Q_1}{Q_2} = F(T_1, T_2);$$

$$\frac{Q_2}{Q_3} = F(T_2, T_3);$$

But

$$\frac{Q_1}{Q_3} = \frac{Q_1 Q_2}{Q_2 Q_3}.$$

Hence

$$\underbrace{F(T_1,T_3)}_{\text{Not a fuction of } T_2} = \underbrace{F(T_1,T_2) \times F(T_2,T_3)}_{\text{Cannot be a function of } T_2}.$$

We thus conclude that $F(T_1, T_2)$ has the form $f(T_1)/f(T_2)$, and similarly $F(T_2, T_3) = f(T_2)/f(T_3)$. The ratio of the heat exchanged is therefore

$$\frac{Q_1}{Q_3} = F(T_1, T_3) = \frac{f(T_1)}{f(T_3)}.$$

In general,

$$\frac{Q_H}{Q_L} = \frac{f(T_H)}{f(T_L)},$$

so that the ratio of the heat exchanged is a function of the temperature. We could choose any function that is monotonic, and one choice is the simplest: $f(T) = T$. This is the thermodynamic scale of temperature, $Q_H/Q_L = T_H/T_L$. The temperature defined in this manner is the same as that for the ideal gas; the thermodynamic temperature scale and the ideal gas scale are equivalent.

REPRESENTATION OF THERMODYNAMIC PROCESSES IN COORDINATES

It is often useful to plot the thermodynamic state transitions and the cycles in terms of temperature (or enthalpy) and entropy, T, S, rather than P, V. The maximum temperature is often the constraint on the process and the enthalpy changes show the work done or heat received directly, so that plotting in terms of these variables provides insight into the process.

A Carnot cycle is shown below in these coordinates, in which it is a rectangle, with two horizontal, constant temperature legs. The other two legs are reversible and adiabatic, hence isentropic ($dS = dQ_{rev}/T = 0$), and therefore vertical in T-s coordinates.

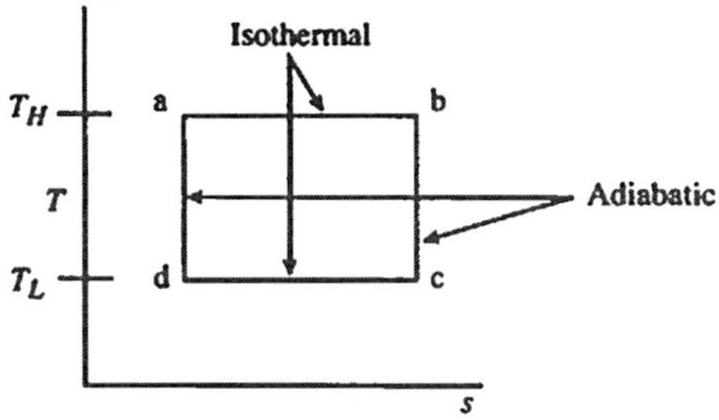

Fig. Carnot Cycle in Coordinates

If the cycle is traversed clockwise, the heat added is

$$\text{Heat added: } Q_H = Q_H = \int_a^b TdS = T_H(S_b - S_a) = T_H \Delta S.$$

The heat rejected (from c to d) has magnitude $|Q_L = T_L \Delta S|$.

The work done by the cycle can be found using the first law for a reversible process:

$$\begin{aligned} dU &= dQ - dW \\ &= TdS - dW \end{aligned}$$

(This form isonly true for a reversible process.)

We can integrate this last expression around the closed path traced out by the cycle:

$$\oint dU = \oint TdS - \oint dW.$$

However dU is an exact differential and its integral around a closed contour is zero:

$$0 = \oint TdS - \oint dW.$$

The work done by the cycle, which is represented by the term $\oint dW$, is equal to $\oint TdS$, the area enclosed by the closed contour in the - plane. This area represents the difference between the heat absorbed ($\oint TdS$ at the high temperature) and the heat rejected ($\oint TdS$ at the low temperature). Finding the work done through evaluation of is an alternative to computation of the work in a reversible cycle from $\oint PdV$. Finally, although we have carried out the discussion in terms of the entropy, S, all of the arguments carry over to the specific entropy, s; the work of the reversible cycle per unit mass is given by $\oint Tds$.

BRAYTON CYCLE IN COORDINATES

The Brayton cycle has two reversible adiabatic (i.e., isentropic) legs and two reversible, constant pressure heat exchange legs. The former are vertical, but we need to define the shape of the latter. For an ideal gas, changes in specific enthalpy are related to changes in temperature by $dh = c_p dT$, so the shape of the cycle in an h - s plane is the same as in a T - s plane, with a scale factor of c_p between the two. This suggests that a place to start is with the combined first and second law, which relates changes in enthalpy, entropy, and pressure:

$$dh = Tds + \frac{dP}{\rho}.$$

On constant pressure curves $dP = 0$ and $dh = Tds$. The quantity desired is the derivative of temperature, T, with respect to entropy, , at constant pressure: $(\partial T/\partial s)_p$. From the combined first and second law, and the relation between dh and dT, this is

$$\left(\frac{\partial T}{\partial s}\right)_p = \frac{T}{c_p}.$$

The derivative is the slope of the constant pressure legs of the Brayton cycle on a T-s plane. For a given ideal gas (specific c_p) the slope is positive and increases as T.

We can also plot the Brayton cycle in an h-s plane. This has advantages because changes in enthalpy directly show the work of the compressor and turbine and the heat added and rejected. The slope of the constant pressure legs in the h-s plane is $(\partial h/\partial s)_p = T$.

Note that the similarity in the shapes of the cycles in *T-s* and *h-s* planes is true for ideal gases only. As we will see when we examine two-phase cycles, the shapes look quite different in these two planes when the medium is not an ideal gas.

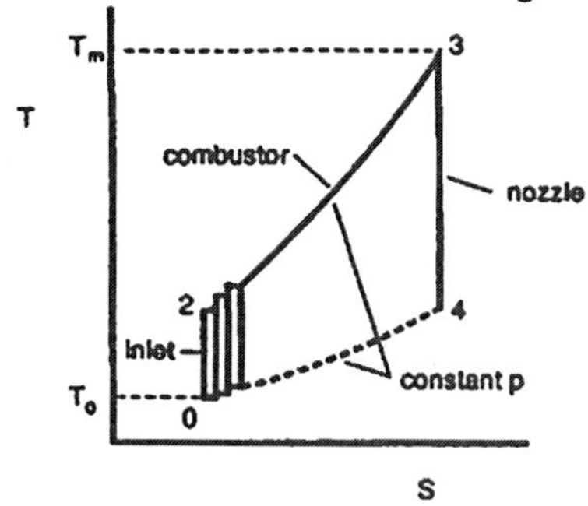

Fig. Ideal Brayton Cycle as Composed of Many Elementary Carnot Cycles

Plotting the cycle in *T-s* coordinates also allows another way to address the evaluation of the Brayton cycle efficiency which gives insight into the relations between Carnot cycle efficiency and efficiency of other cycles. We can break up the Brayton cycle into many small Carnot cycles. The "*i*th" Carnot cycle has an efficiency of

$$\eta C_i = 1 - \left(\frac{T_{\text{low},i}}{T_{\text{high},i}}\right),$$

where the indicated lower temperature is the heat rejection temperature for that elementary cycle and the higher temperature is the heat absorption temperature for that cycle. The upper and lower curves of the Brayton cycle, however, have constant pressure. All of the elementary Carnot cycles therefore have the same pressure ratio:

$$\frac{P(T_{\text{high}})}{P(T_{\text{low}})} = PR = \text{constant} \quad \text{(the same for all cycles).}$$

From the isentropic relations for an ideal gas, we know that pressure ratio, *PR*, and temperature ratio, *TR*, are related by: $PR^{(\gamma-1)/\gamma} = TR$.

The temperature ratios ($T_{\text{low},\,i}/T_{\text{high},\,i}$) of any elementary cycle "*i*" are therefore the same and each of the elementary cycles has the same thermal efficiency. We only need to find the temperature ratio across any one of the cycles to find what the efficiency is. We know that the temperature ratio of the first elementary cycle is the ratio of compressor exit temperature to engine entry (atmospheric for an aircraft engine) temperature, T_2/T_0. If the efficiency of all the elementary cycles has this value, the efficiency of the overall Brayton cycle (which is composed of the elementary cycles) must also have this value. Thus, as previously,

$$\eta_{\text{Brayton}} = 1 - \frac{T_{\text{inlet}}}{T_{\text{compressor}}}.$$

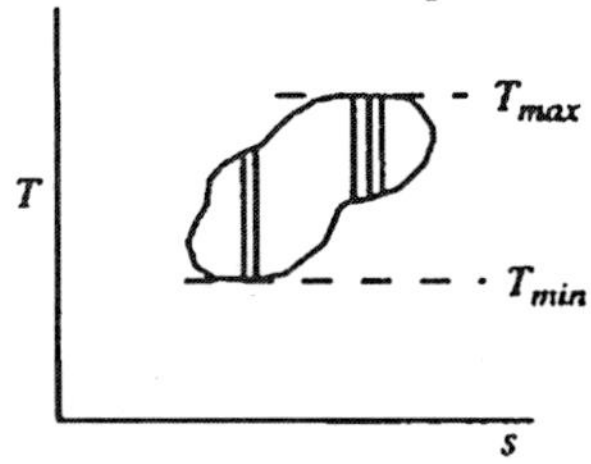

Fig. Arbitrary Cycle Operating between T_{min}, T_{max}

A benefit of this view of efficiency is that it allows us a way to comment on the efficiency of any thermodynamic cycle. Consider the cycle, which operates between some maximum and minimum temperatures.

We can break it up into small Carnot cycles and evaluate the efficiency of each. It can be seen that the efficiency of any of the small cycles drawn will be less than the efficiency of a Carnot cycle between T_{max} and T_{min}.

This graphical argument shows that the efficiency of any other thermodynamic cycle operating between these maximum and minimum temperatures has an efficiency less than that of a Carnot cycle.

NET WORK PER UNIT MASS FLOW IN A BRAYTON CYCLE

We found the net work of a Brayton cycle in terms of heat transfer. Now that we have defined entropy, we can reexamine the net work using an enthalpy-entropy (*h* - *s*) diagram. The net mechanical work of the cycle is given by:

Net mechanical work/unit mass = $w_{turbine} - w_{compressor}$, where, by the first law,

$$w_{compressor} = -\Delta h_{03} = \Delta h_{comp}$$

$$w_{turbine} = \Delta h_{45} = -\Delta h_{turb}.$$

If kinetic energy changes across the compressor and turbine are neglected, the temperature ratio, *TR*, across the compressor and turbine is related to the enthalpy changes:

$$TR - 1 = \frac{\Delta h_{comp}}{h_0} = \frac{|\Delta h_{turb}|}{h_5},$$

$$\Delta h_{turb} = -\Delta h_{comp}\frac{h_5}{h_0}.$$

The net work is thus

$$\text{net work } = \Delta h_{comp}\left(\frac{h_5}{h_0} - 1\right).$$

The turbine work is greater than the work needed to drive the compressor, as is evident on the (h - s) diagram.

IRREVERSIBILITY, ENTROPY CHANGES, AND "LOST WORK"

Consider a system in contact with a heat reservoir during a reversible process. If there is heat Q absorbed by the reservoir at temperature T, the change in entropy of the reservoir is $\Delta S = QT$. In general, reversible processes are accompanied by heat exchanges that occur at different temperatures. To analyze these, we can visualize a sequence of heat reservoirs at different temperatures so that during any infinitesimal portion of the cycle there will not be any heat transferred over a finite temperature difference.

During any infinitesimal portion, heat dQ_{rev} will be transferred between the system and one of the reservoirs which is at T. If dQ_{rev} is absorbed by the system, the entropy change of the system is

$$dS_{system} = \frac{dQ_{rev}}{T}.$$

The entropy change of the reservoir is

$$dS_{\text{reservoir}} = \frac{dQ_{\text{rev}}}{T}.$$

The total entropy change of system plus surroundings is

$$dS_{\text{total}} = dS_{\text{system}} + dS_{\text{reservoir}} = 0.$$

This is also true if there is a quantity of heat rejected by the system.

The conclusion is that for a reversible process, no change occurs in the total entropy produced, i.e., the entropy of the system plus the entropy of the surroundings: $\Delta S_{\text{total}} = 0$.

Fig. Irreversible and Reversible State Changes

We now carry out the same type of analysis for an irreversible process, which takes the system between the same specified states as in the reversible process with and denoting the irreversible and reversible processes. In the irreversible process, the system receives heat dQ and does work dW. The change in internal energy for the irreversible process is

$$dU = dQ - dW \text{ (Always true — first law).}$$

For the reversible process

$$dU = TdS - dW_{\text{rev}}.$$

Because the state change is the same in the two processes (we specified that it was), the change in internal energy is the same. Equating the changes in internal energy in the above two expressions yields

$$dQ_{\text{actual}} - dW_{\text{actual}} = TdS - dW_{\text{rev}}.$$

The subscript "actual" refers to the actual process (which is irreversible). The entropy change associated with the state change is

$$dS = \frac{dQ_{\text{actual}}}{T} + \frac{1}{T}\left[dW_{\text{rev}} - dW_{\text{actual}}\right].$$

If the process is not reversible, we obtain less work than in a reversible process, $dW_{\text{actual}} < dW_{\text{rev}}$, so that for the irreversible process,

$$dS > \frac{dQ_{\text{actual}}}{T}.$$

There is no equality between the entropy change dS and the quantity dQ/T for an irreversible process. The equality is only applicable for a reversible process.

The change in entropy for any process that leads to a transformation between an initial state "a" and a final state "b" is therefore

$$\Delta S = S_b - S_a \geq \int_a^b \frac{dQ_{\text{actual}}}{T},$$

where dQ_{actual} is the heat exchanged in the actual process. The equality only applies to a reversible process.

The difference $dW_{\text{rev}} - dW_{\text{actual}}$ represents work we could have obtained, but did not. It is referred to as lost work and denoted by W_{lost}. In terms of this quantity we can write,

$$dS = \frac{dQ_{\text{actual}}}{T} + \frac{dW_{\text{lost}}}{T}.$$

The content of Equation is that the entropy of a system can be altered in two ways: (i) through heat exchange and (ii) through irreversibilities. The lost work (dW_{lost} in Equation) is always greater than zero, so the only way to decrease the entropy of a system is through heat transfer.

To apply the second law we consider the total entropy change (system plus surroundings). If the surroundings are a reservoir at temperature T, with which the system exchanges heat,

$$dS_{\text{reser voir}} (= dS_{\text{surroundings}}) = \frac{dQ_{\text{actual}}}{T}.$$

The total entropy change is

$$d_{\text{Stotal}} = dS_{\text{system}} + dS_{\text{surroundings}} = \left(\frac{dQ_{\text{actual}}}{T} + \frac{dW_{\text{lost}}}{T}\right) - \frac{dQ_{\text{actual}}}{T}.$$

$$dS_{\text{total}} = \frac{dW_{\text{lost}}}{T} \geq 0.$$

The quantity (dW_{lost}/T) is the entropy generated due to irreversibility.

Yet another way to state the distinction we are making is

$$dS_{\text{system}} = dS_{\text{from heat transfer}} + dS_{\text{generated due to irreversible processes}}$$
$$= dS_{\text{heat transfer}} + dS_{\text{Gen}}.$$

The lost work is also called dissipation and noted $d\Phi$. Using this notation, the infinitesimal entropy change of the system becomes:

$$dS_{\text{system}} = dS_{\text{heat transfer}} + \frac{d\Phi}{T}$$

or

$$Td_{\text{Ssystem}} = dQ + d\Phi.$$

Equation can also be written as a rate equation,

$$\frac{dS}{dt} = \dot{S} = \dot{S}_{\text{heat transfer}} + \dot{S}_{\text{Gen}}.$$

Either of Equations can be interpreted to mean that the entropy of the system, S, is affected by two factors: the flow of heat Q and the appearance of additional entropy, denoted by dS_{Gen}, due to irreversibility. This additional entropy is zero when the process is reversible and always positive when the process is irreversible.

Thus, one can say that the system develops sources which create entropy during an irreversible process. The second law asserts that sinks of entropy are impossible in nature, which is a more graphic way of saying that dS_{Gen} and $\dot{S}_{\text{Gen}}$ are positive definite (always greater than zero), or zero in the special case of reversible processes.

The term

$$\dot{S}_{\text{heat transfer}} \left(= \frac{1}{T}\frac{dQ}{dt}, \quad \text{or} \quad \frac{\dot{Q}}{T} \right)$$

which is associated with heat transfer to the system, can be interpreted as a flux of entropy. The boundary is crossed by heat and the ratio of this heat flux to temperature can be defined as a flux of entropy. There are no restrictions on the sign of this quantity, and we can say that this flux either contributes towards, or drains away, the system's entropy.

During a reversible process, only this flux can affect the entropy of the system. This terminology suggests that we interpret entropy as a kind of weightless fluid, whose quantity is conserved (like that of matter) during a reversible process.

During an irreversible process, however, this fluid is not conserved; it cannot disappear, but rather is created by sources throughout the system.

While this interpretation should not be taken too literally, it provides an easy mode of expression and is in the same category of concepts such as those associated with the phrases "flux of energy" or "sources of heat." In fluid mechanics, This graphic language is very effective and there should be no objections to copying it in thermodynamics.

ENTROPY AND UNAVAILABLE ENERGY

Consider a system consisting of a heat reservoir at in surroundings (the atmosphere) at T_2. The surroundings are equivalent to a second reservoir at T_0. For an amount of heat, Q, transferred from the reservoir, the maximum work we could derive is Q times the thermal efficiency of a Carnot cycle operated between these two temperatures:

$$\text{Maximum work we could} = \text{Wmax} = Q\left(1-\frac{T_0}{T_2}\right).$$

Only part of the heat transferred can be turned into work, in other words only part of the heat energy is available to be used as work.

Suppose we transferred the same amount of heat from the reservoir directly to another reservoir at a temperature $T_1 < T_2$. The maximum work available from the quantity of heat, Q, before the transfer to the reservoir at T_1 is

$$W_{\max,\, T_2,T_0} = Q\left(1-\frac{T_0}{T_2}\right); \text{ (Maximum work between } T_2, T_0).$$

The maximum amount of work available after the transfer to the reservoir at T_1 is

$$W_{\max,\, T_1,T_0} = Q\left(1-\frac{T_0}{T_2}\right); \text{ (Maximum work between } T_1, T_0).$$

There is an amount of energy that could have been converted to work prior to the irreversible heat transfer process of magnitude E',

$$E' = Q\left[1-\left(\frac{T_0}{T_2}\right)-\left(1-\frac{T_0}{T_1}\right)\right] = Q\left[\frac{T_0}{T_1}-\frac{T_0}{T_2}\right],$$

or $$E' = T_0\left[\frac{Q}{T_1}-\frac{Q}{T_2}\right].$$

However, Q/T_1 is the entropy gain of the reservoir at and $T_1(-Q/T_2)$ is the entropy decrease of the reservoir at T_2. The amount of energy, E', that could have been converted to work (but now cannot be) can therefore be written in terms of entropy changes and the temperature of the surroundings as

$$E' = T_0(\Delta S_{\text{reservoir at } T_1} + \Delta S_{\text{reservoir at } T_2})$$

$$= T_0\ \Delta S_{\text{irreversible heat transfer process}}$$

E' = ''Lost work,'' orenergy which isnolonger avialable to do work.

The situation just described is a special case of an important principle concerning entropy changes, irreversibility and the loss of capability to do work. We thus now develop it in a more general fashion, considering an arbitrary system undergoing an irreversible state change, which transfers heat to the surroundings, which can be assumed to be at constant temperature, T_0. The change in internal energy of the system during the state change is $\Delta U = Q - W$. The change in entropy of the surroundings is (with Q the heat transfer to the system)

$$\Delta S_{\text{surroundings}} = -\frac{Q}{T_0}.$$

Now consider restoring the system to the initial state by a reversible process. To do this we need to do work, W_{rev}, on the system and extract from the system a quantity of heat, Q_{rev}. (In "undoing" the free expansion process.) The change in internal energy is (with the quantities Q_{rev} and W_{rev} both regarded, in this example, as positive for work done by the surroundings and heat given to the surroundings).

$$\Delta U_{rev} = -Q_{rev} + W_{rev}.$$

In this reversible process, the entropy of the surroundings is changed by

$$\Delta S_{surroundings} = \frac{Q_{rev}}{T}.$$

For the combined changes (the irreversible state change and the reversible state change back to the initial state), the energy change is zero because the energy is a function of state,

$$\Delta U_{rev} + \Delta U = 0 = Q - W + (-Q_{rev} + W_{rev}).$$

Thus,

$$Q_{rev} - Q = W_{rev} - \mathrm{W}.$$

For the system, the overall entropy change for the combined process is zero, because the entropy is a function of state,

$$\Delta S_{system,combined\ process} = \Delta S_{irreversible\ process} + \Delta S_{reversible\ process} = 0.$$

The total entropy change is thus only reflected in the entropy change of the surroundings:

$$\Delta S_{total} = \Delta S_{surroundings}.$$

The surroundings can be considered a constant temperature heat reservoir and their entropy change is given by

$$\Delta S_{total} = \frac{(Q_{rev} - Q)}{T_0}.$$

We also know that the total entropy change, for system plus surroundings is,

$$\Delta S_{total} = \left[\Delta S_{irreversible\ process} + \cancel{\Delta S}^{0}_{reversible\ process}\right]_{system\ and\ surroundings}.$$

The total entropy change is associated only with the irreversible process and is related to the work in the two processes by

$$\Delta S_{total} = \frac{(W_{rev} - W)}{T_0}.$$

The quantity $W_{rev} - W$ represents the extra work required to restore the system to the original state. If the process were reversible, we would not have needed any extra work to do this. It represents a quantity of work that is now unavailable because of the irreversibility. The quantity W_{rev} can also be interpreted as the work that the system would have done if the original process were reversible. From either of these perspectives we can identify (W_{rev} – W) as the quantity we denoted previously as E', representing lost work. The lost work in any irreversible process can therefore be related to the total entropy change (system plus surroundings) and the temperature of the surroundings by

$$\text{Lost work} = W_{rev} - W = T_0 \Delta S_{total}.$$

To summarize the results of the above arguments for processes where heat can be exchanged with the surroundings at:

- $W_{rev} - W$ represents the difference between work we actually obtained and work that would be done during a reversible state change. It is the extra work that would be needed to restore the system to its initial state.
- For a reversible process, $W_{rev} = W$; $\Delta S_{total} = 0$.
- For an irreversible process, $W_{rev} > W$; $\Delta S_{total} > 0$.
- $(W_{rev} - W) = E' = T_0\ \Delta S_{total}$ is the energy that becomes unavailable for work during an irreversible process.

EXAMPLES OF LOST WORK IN ENGINEERING PROCESSES

1. Lost work in Adiabatic Throttling: Entropy and Stagnation Pressure Changes

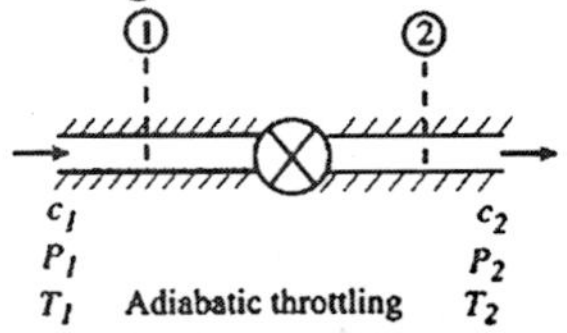

Fig. Adiabatic Throttling

A process we have encountered before is adiabatic throttling of a gas, by a valve or other device. The velocity is denoted by c. There is no shaft work and no heat transfer and the flow is steady. Under these conditions we can use the first law for a control volume (the Steady Flow Energy Equation) to make a statement about the conditions upstream and downstream of the valve:

$$h_1 + \frac{c_1^2}{2} = h_2 + \frac{c_2^2}{2} = h_t,$$

where h_t is the stagnation enthalpy, corresponding to a (possibly fictitious) state with zero velocity. The stagnation enthalpy is the same at stations 1 and 2 if $Q = W = 0$, even if the flow processes are not reversible. For a perfect gas with constant specific heats, $h = c_p T$ and $h_t = c_p T_t$. The relation between the static and stagnation temperatures is:

$$\frac{Tt}{T} = 1 + \frac{c^2}{2_p T} = 1 + \underbrace{\frac{(\gamma - 1)}{2\gamma RT}}_{a^2} = 1 + \frac{(\gamma - 1)c^2}{2a^2},$$

$$\frac{Tt}{T} = 1 + \left(\frac{\gamma - 1}{2}\right) M^2,$$

where a is the speed of sound and M is the Mach number, $M = c/a$. In deriving this result, use has only been made of the first law, the equation of state, the speed of sound, and the definition of the Mach number. Nothing has yet been specified about whether the process of stagnating the fluid is reversible or irreversible.

When we define the stagnation pressure, however, we do it with respect to isentropic deceleration to the zero velocity state. For an isentropic process

$$\frac{P_2}{P_1} = \left(\frac{T_2}{T_1}\right)^{\gamma/(\gamma-1)}.$$

The relation between static and stagnation pressures is

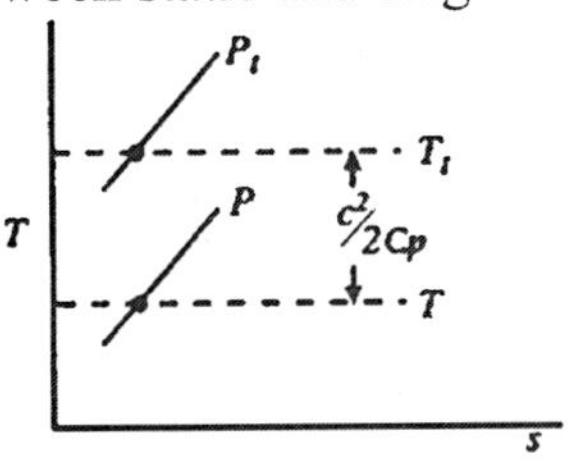

Fig. Static and Stagnation Pressures and Temperatures

The stagnation state is defined by P_t, T_t. In addition, $S_{\text{stagnation state}}$. The static and stagnation states are shown in T - s coordinates.

Stagnation pressure is a key variable in propulsion and power systems. To see why, we examine the relation between stagnation pressure, stagnation temperature, and entropy. The form of the combined first and second law that uses enthalpy is

$$Tds = dh - \frac{1}{\rho}dP.$$

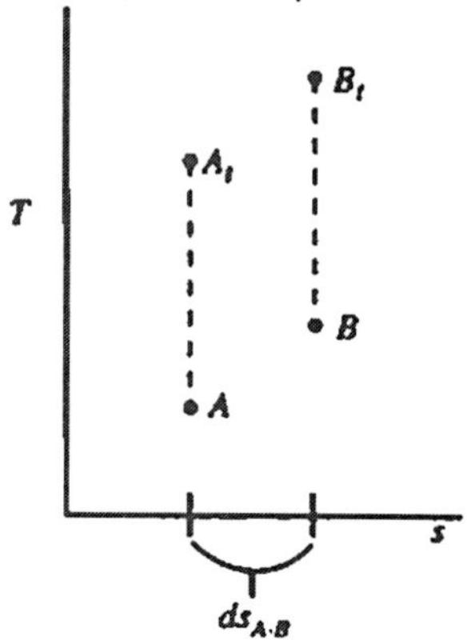

Fig. Stagnation and Static States

This holds for small changes between any thermodynamic states and we can apply it to a situation in which we consider differences between stagnation states, say one state having properties (P_t, P_t)and the other having properties ($T_t + dT_t$, $P_t + dP_t$). The corresponding static states are also indicated. Because the entropy is the same at static and stagnation conditions, needs no subscript. Writing in terms of stagnation conditions yields

$$ds = \frac{c_p dT_t}{T_t} - \frac{1}{\rho_t T_t}dP_t = \frac{c_p dT_t}{T_t} - \frac{R}{P_t}dp_t.$$

Both sides of the above are perfect differentials and can be integrated as

$$\frac{\Delta s}{R} = \frac{\gamma}{\gamma - 1} = \left(\frac{T_{t2}}{T_{t1}}\right) - \ln\left(\frac{P_{t2}}{P_{t1}}\right).$$

For a process with $Q = W = 0$, the stagnation enthalpy, and hence the stagnation temperature, is constant. In this situation, the stagnation pressure is related directly to the entropy as,

$$\frac{\Delta s}{R} = -\ln\left(\frac{P_{t2}}{P_{t1}}\right).$$

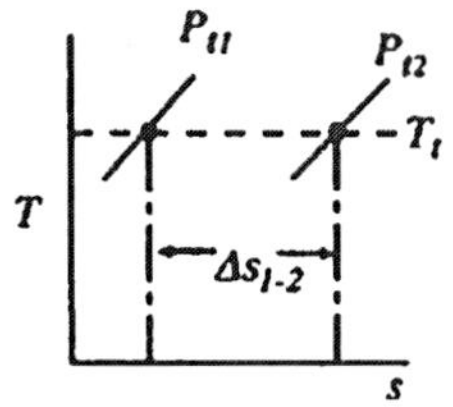

Fig. Losses reflected in changes in stagnation pressure when

This relation on a T-s diagram. We have seen that the entropy is related to the loss, or irreversibility. The stagnation pressure plays the role of an indicator of loss if the stagnation temperature is constant. The utility is that it is the stagnation pressure (and temperature) which are directly measured, not the entropy. The throttling process is a representation of flow through inlets, nozzles, stationary turbomachinery blades, and the use of stagnation pressure as a measure of loss is a practice that has widespread application.

Equation can be put in several useful approximate forms. First, we note that for aerospace applications we are (hopefully!) concerned with low loss devices, so that the stagnation pressure change is small compared to the inlet level of stagnation pressure,

$$\frac{\Delta P_t}{P_{t1}} = \frac{(P_{t1} - P_{t2})}{P_{t1}} << 1.$$

Expanding the logarithm (using $\ln(1 - x) \cong -x + ...$),

$$\ln\left(\frac{P_{t2}}{P_{t1}}\right) = \ln\left(1 - \frac{\Delta P_t}{P_{t1}}\right) \approx -\frac{\Delta P_t}{P_{t1}},$$

or

$$\frac{\Delta s}{R} \approx \frac{\Delta P_t}{P_{t1}}.$$

Another useful form is obtained by dividing both sides by $c^2/2$ and taking the limiting forms of the expression for stagnation pressure in the limit of low Mach number (M << 1). Doing this, we find:

$$\frac{T\Delta s}{c^2/2} \cong \frac{\Delta P_t}{\rho c^2/2}.$$

The quantity on the right can be interpreted as the change in the "Bernoulli constant" for incompressible (low Mach number) flow. The quantity on the left is a non-dimensional entropy change parameter, with the term $T\Delta s$ now representing the loss of mechanical energy associated with the change in stagnation pressure.

To summarize: For many applications the stagnation temperature is constant and the change in stagnation pressure is a direct measure of the entropy increase.

Stagnation pressure is the quantity that is actually measured so that linking it to entropy (which is not measured) is useful.

We can regard the throttling process as a "free expansion" at constant temperature T_{t1} from the initial stagnation pressure to the final stagnation pressure. We thus know that, for the process, the work we need to do to bring the gas back to the initial state is $T_t\Delta s$, which is the "lost work" per unit mass.

2. Adiabatic Efficiency of a Propulsion System Component (Turbine)

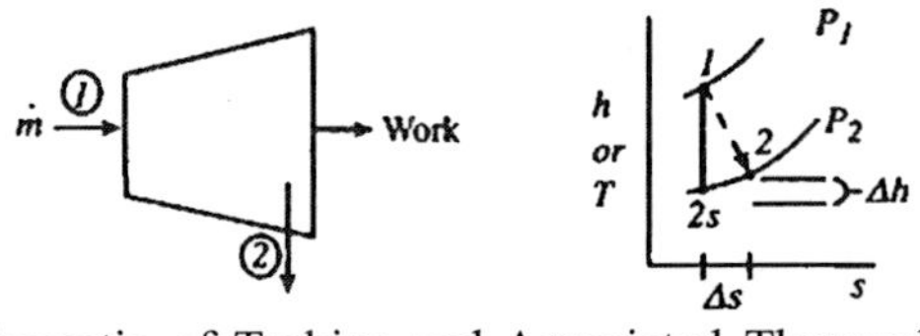

Fig. Schematic of Turbine and Associated Thermodynamic Representation in *h-s* Coordinates

A schematic of a turbine and the accompanying thermodynamic. There is a pressure and temperature drop through the turbine and it produces work. There is no heat transfer so the expressions that describe the overall shaft work and the shaft work per unit mass are

$$\dot{m}(h_{t2} - h_{t1}) = \dot{W}_{\text{shaft}}$$

$$(h_{t2} - h_{t1}) = w_{\text{shaft}}.$$

If the difference in the kinetic energy at inlet and outlet can be neglected, Equation reduces to

$$(h_2 - h_1) = w_{\text{shaft}}.$$

The adiabatic efficiency of the turbine is defined as

$$\eta_{ad} = \left[\frac{\text{actual work}}{\text{ideal work } (\Delta s = 0)}\right]_{\text{for a given pressure ratio}}.$$

The performance of the turbine can be represented in an *h-s* plane (similar to a *T- s* plane for a perfect gas with constant specific heats). From the figure the adiabatic efficiency is

$$\eta_{ad} = \frac{h_1 - h_2}{h_1 - h_{2s}} = \frac{h_1 - h_{2s} - (h_2 - h_{2s})}{h_1 - h_{2s}}.$$

The adiabatic efficiency can therefore be written as

$$\eta_{ad} = 1 - \left(\frac{\Delta h}{\text{Ideal work}}\right).$$

The non-dimensional term (Δh/Ideal work) represents the departure from isentropic (reversible) processes and hence a loss. The quantity Δh is the enthalpy difference for two states along a constant pressure line. From the combined first and second laws, for a constant pressure process, small changes in enthalpy are related to the entropy change by $Tds = dh$, or approximately,

$$T_2\Delta s = \Delta h.$$

The adiabatic efficiency can thus be approximated as

$$\eta_{ad} = 1 - \frac{T_2\Delta s}{h_1 - h_{2s}} = 1 - \left(\frac{\text{Lost work}}{\text{Ideal work}}\right).$$

The quantity $T\Delta s$ represents a useful figure of merit for fluid machinery inefficiency due to irreversibility.

3. Isothermal Expansion with Friction

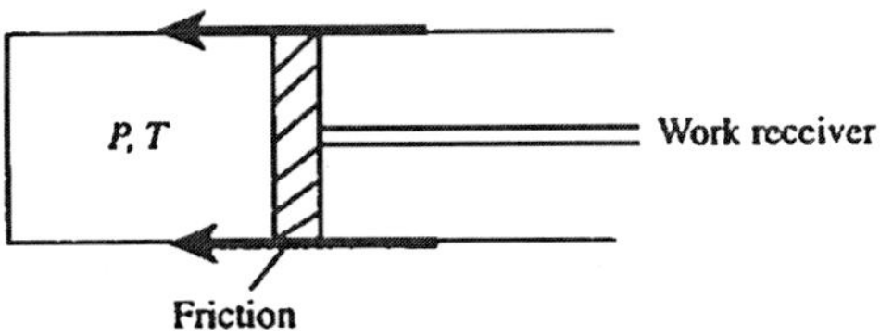

Fig. Isothermal Expansion with Friction

In a more general look at the isothermal expansion, we now

drop the restriction to frictionless processes work is done to overcome friction. If the kinetic energy of the piston is negligible, a balance of forces tells us that

$$W_{\text{system on piston}} = W_{\text{done by friction}} + W_{\text{received}}.$$

During the expansion, the piston and the walls of the container will heat up because of the friction. The heat will be (eventually) transferred to the atmosphere; all frictional work ends up as heat transferred to the surrounding atmosphere.

$$W_{\text{friction}} = Q_{friction}.$$

The amount of heat transferred to the atmosphere due to the frictional work only is thus,

$$Q_{\text{friction}} = \underbrace{W_{\text{system on piston}}}_{\text{Work produced}} - \underbrace{W_{\text{received}}}_{\text{Work produced}}.$$

The entropy change of the atmosphere (considered as a heat reservoir) due to the frictional work is

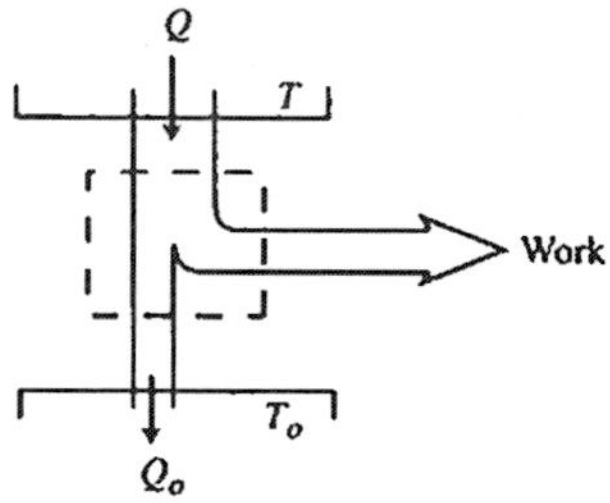

The engine operates in a cycle and the entropy change for the complete cycle is zero (because entropy is a state variable). Therefore,

$$\Delta S = 0 + \underbrace{\Delta_{\text{heat source}} + \Delta S_{\text{heat sink}}}_{\Delta S_{\text{surroundings}}}.$$

The total entropy change is

$$\Delta S_{\text{total}} = \Delta S_{\text{heat source}} + \Delta S_{\text{heat sink}} = -\frac{Q}{T} + \frac{Q_0}{T_0}.$$

Suppose we had an ideal reversible engine working between these same two temperatures, which extracted the same amount of heat, Q, from the high temperature reservoir, and rejected heat of magnitude $Q_{0,\text{ rev}}$ to the low temperature reservoir. The work done by this reversible engine is

$$W_{rev} = Q - Q_{0,\, rev}.$$

For the reversible engine the total entropy change over a cycle is

$$\Delta S_{total} = \Delta S_{heat\ source} + \Delta S_{heat\ sink} = -\frac{Q}{T} + \frac{Q_{0,\, rev}}{T_0} = 0.$$

Combining the expressions for work and for the entropy changes,

$$Q_0 = Q_{0,rev} + W_{rev} - W.$$

The entropy change for the irreversible cycle can therefore be written as

$$\Delta S_{total} = \underbrace{-\frac{Q}{T} + \frac{Q_{0,\, rev}}{T_0}}_{=0} + \frac{W_{rev} - W}{T_0}$$

The difference in work that the two cycles produce is proportional to the entropy that is generated during the cycle:

$$T_0 \Delta S_{total} = W_{rev} - W.$$

The second law states that the total entropy generated is greater than zero for an irreversible process, so that the reversible work is greater than the actual work of the irreversible cycle.

An "engine effectiveness," E_{engine}, can be defined as the ratio of the actual work obtained divided by the work that would have been delivered by a reversible engine operating between the two temperatures T and T_0:

$$E_{engine} = \frac{\eta_{engine}}{\eta_{reversible\ engine}}$$

$$= \frac{W}{W_{rev}} = \frac{\text{Actual work obtained}}{\text{Work that \textit{would be delivered} by a reversible cycle between } T,\ T_0}$$

$$E_{engine} = \frac{W_{rev} - T_0 \Delta S_{total}}{W\ rev} = 1 - \frac{T_0 \Delta S_{total}}{W_{rev}}.$$

The departure from a reversible process is directly reflected in the entropy change and the decrease in engine effectiveness.

Propulsive Power and Entropy Flux: The final example in this section combines a number of ideas presented in this subject and in Unified in the development of a relation between entropy generation and power needed to propel a vehicle.

An aerodynamic shape (airfoil) moving through the atmosphere at a constant velocity. A coordinate system fixed to the vehicle has been adopted so that we see the airfoil as fixed and the air far away from the airfoil moving at a velocity c_0. Streamlines of the flow have been sketched, as has the velocity distribution at station "0" far upstream and station "d" far downstream.

The airfoil has a wake, which mixes with the surrounding air and grows in the downstream direction. The extent of the wake is also indicated. Because of the lower velocity in the wake the area between the stream surfaces is larger downstream than upstream.

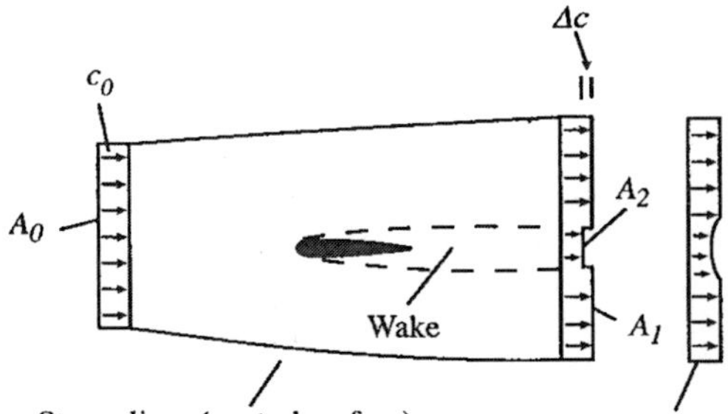

Fig. Airfoil with Wake and Control Volume for Analysis of Propulsive power requirement

We use a control volume description and take the control surface to be defined by the two stream surfaces and two planes at station 0 and station d. This is useful in simplifying the analysis because there is no flow across the stream surfaces. The area of the downstream plane control surface is broken into A_1, which is the area outside the wake and A_2, which is the area occupied by wake fluid, i.e., fluid that has suffered viscous losses. The control surface is also taken far enough away from the vehicle so that the static pressure can be considered uniform. For fluid which is not in the wake (no viscous forces), the momentum equation is

$$cdc = -\, dP/\rho.$$

Uniform static pressure therefore implies uniform velocity, so that on A_1 the velocity is equal to the upstream value, c_0. The downstream velocity profile is actually continuous, as indicated. It is approximated in the analysis as a step change to make the algebra a bit simpler.

The equation expressing mass conservation for the control volume is

$$\rho_0 A_0 c_0 = \rho_0 A_1 c_0 + \rho_2 A_2 c_2.$$

The vertical face of the control surface is far downstream of the body. By this station, the wake fluid has had much time to mix and the velocity in the wake is close to the free stream value, c_0. We can thus write,

$$\text{wake velocity} = c_2 = (c_0 - \Delta c); \quad \Delta c/c_0 << 1.$$

(We chose our control surface so the condition $\Delta c/c_0 << 1$ was upheld.)

The integral momentum equation (control volume form of the momentum equation) can be used to find the drag on the vehicle:

$$\rho_0 A_0 c_0^2 = -\text{Drag} + \rho_0 A_1 c_0^2 + \rho_2 A_2 c_2^2.$$

There is no pressure contribution in Equation because the static pressure on the control surface is uniform. Using the form given for the wake velocity and expanding the terms in the momentum equation we obtain

$$\rho_0 A_0 c_0^2 = -\text{Drag} + \rho_0 A_1 c_0^2 + \rho_2 A_2 [c_2^2. - 2c_0 \Delta c + (\Delta c)^2].$$

The last term in the right hand side of the momentum equation, $\rho_2 A_2 (\Delta c)^2$, is small by virtue of the choice of control surface and we can neglect it. Doing this and grouping the terms on the right hand side of Equation in a different manner, we have

$$c_0[\rho_0 A_0 c_0] = c_0[\rho_0 A_1 c_0 + \rho_2 A_2 (c_0 - \Delta c)] + \{-\text{Drag} - \rho_2 A_2 c_0 \Delta c\}.$$

The terms in the square brackets on both sides of this equation are the continuity equation multiplied by c_0. They thus sum to zero leaving the curly bracketed terms as

$$\text{Drag} = -\rho_0 A_0 c_0 \Delta c.$$

The wake mass flow is $\rho_2 A_2 c_2 = \rho_2 A_2 (c_0 - \Delta c)$. All this flow has a velocity defect (compared to the free stream) of Δc, so that the defect in flux of momentum (the mass flow in the wake times the velocity defect) is, to first order in Δc,

Momentum defect in wake = $-\rho_2 A_2 c_0 \Delta c$ = Drag.

The combined first and second law gives us a means of relating the entropy and velocity:

$$Tds = dh - dP/\rho.$$

The pressure is uniform ($dP = 0$) at the downstream station. There is no net shaft work or heat transfer to the wake so that the mass flux of stagnation enthalpy is constant. We can also approximate that the condition of constant stagnation enthalpy holds locally on all streamlines. Applying both of these to the combined first and second law yields

$$Tds = dh_t - cdc.$$

For the present situation, $dh_t = 0$; $cdc = c_0\Delta c$, so that

$$T_0\Delta s = -c_0\Delta c.$$

In Equation the upstream temperature is used because differences between wake quantities and upstream quantities are small at the downstream control station. The entropy can be related to the drag as

$$\text{Drag} = \rho_2 A_2 T_0 \Delta s.$$

The quantity $\rho_2 A_2 c_0 \Delta s$ is the entropy flux (mass flux times the entropy increase per unit mass; in the general case we would express this by an integral over the locally varying wake velocity and density). The power needed to propel the vehicle is the product of drag × flight speed, Drag × c_0.

From Equation, this can be related to the entropy flux in the wake to yield a compact expression for the propulsive power needed in terms of the wake entropy flux:

Propulsive power needed = $T_0\,(\rho_0 A_2\, c_0 \Delta s) = T_0$ × Extropy flux wake.

This amount of work is dissipated per unit time in connection with sustaining the vehicle motion. Equation is another demonstration of the relation between lost work and entropy generation, in this case manifested as power that needs to be supplied because of dissipation in the wake.

REVERSIBLE AND IRREVERSIBLE PROCESSES

Entropy

- Entropy is a thermodynamic property that measures the

degree of randomization or disorder at the microscopic level. The natural state of affairs is for entropy to be produced by all processes.

- A macroscopic feature which is associated with entropy production is a loss of ability to do useful work. Energy is degraded to a less useful form, and it is sometimes said that there is a decrease in the availability of energy.
- Entropy is an extensive thermodynamic property. In other words, the entropy of a complex system is the sum of the entropies of its parts.
- The notion that entropy can be produced, but never destroyed, is the second law of thermodynamics.

Reversible and Irreversible Processes

Processes can be classed as reversible or irreversible. The concept of a reversible process is an important one which directly relates to our ability to recognize, evaluate, and reduce irreversibilities in practical engineering processes.

Consider an isolated system. The second law says that any process that would reduce the entropy of the isolated system is impossible. Suppose a process takes place within the isolated system in what we shall call the forward direction. If the change in state of the system is such that the entropy increases for the forward process, then for the backward process (that is, for the reverse change in state) the entropy would decrease. The backward process is therefore impossible, and we therefore say that the forward process is irreversible.

If a process occurs, however, in which the entropy is unchanged by the forward process, then it would also be unchanged by the reverse process. Such a process could go in either direction without contradicting the second law. Processes of this latter type are called reversible. The key idea of a reversible process is that it does not produce any entropy.

Entropy is produced in irreversible processes. All real processes (with the possible exception of superconducting current flows) are in some measure irreversible, though many processes can be analyzed quite adequately by assuming that they are reversible. Some processes that are clearly irreversible include: mixing of two gases, spontaneous combustion, friction, and the transfer of energy as heat from a body at high temperature to a body at low temperature.

Recognition of the irreversibilities in a real process is especially important in engineering. Irreversibility, or departure from the ideal condition of reversibility, reflects an increase in the amount of disorganized energy at the expense of organized energy. The organized energy (such as that of a raised weight) is easily put to practical use; disorganized energy (such as the random motions of the molecules in a gas) requires "straightening out" before it can be used effectively. Further, since we are always somewhat uncertain about the microscopic state, this straightening can never be perfect. Consequently the engineer is constantly striving to reduce irreversibilities in systems, in order to obtain better performance.

EXAMPLES OF REVERSIBLE AND IRREVERSIBLE PROCESSES

Processes that are usually idealized as reversible include:

- Frictionless movement
- Restrained compression or expansion
- Energy transfer as heat due to infinitesimal temperature nonuniformity
- Electric current flow through a zero resistance
- Restrained chemical reaction
- Mixing of two samples of the same substance at the same state.

Processes that are irreversible include:

- Movement with friction
- Unrestrained expansion
- Energy transfer as heat due to large temperature non uniformities
- Electric current flow through a non zero resistance
- Spontaneous chemical reaction
- Mixing of matter of different composition or state.

REAL CYCLE BEHAVIOUR

We will now improve our estimates of cycle performance by including the effects of irreversibility. We will use the Brayton cycle as an example. What are the sources of non-ideal performance and departures from reversibility?

- Losses (entropy production) in the compressor and the turbine.
- Stagnation pressure decrease in the combustor.
- Heat transfer.

We take into account here only irreversibility in the compressor and in the turbine. Because of these irreversibilities, we need more work, Δh_{comp}(the changes in kinetic energy from inlet to exit of the compressor are neglected), to drive the compressor than in the ideal situation. We also get less work, Δh_{turb}, back from the turbine. The consequence, as can be inferred is that the net work from the engine is less than in the cycle with ideal components.

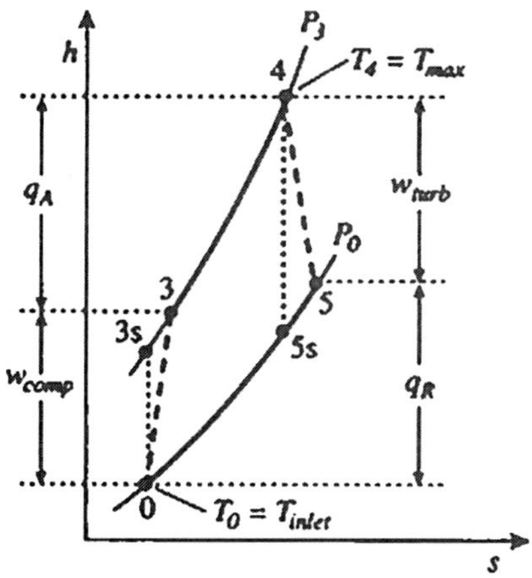

Fig. Gas Turbine Engine (Brayton) Cycle Showing Effect of Departure from Ideal Behaviour in Compressor and Turbine

To develop a quantitative description of the effect of these departures from reversible behaviour, consider a perfect gas with constant specific heats and neglect kinetic energy at the inlet and exit of the turbine and compressor. We define the turbine adiabatic efficiency as

$$\eta_{\text{turb}} = \frac{w_{\text{turb}}^{\text{actual}}}{w_{\text{turb}}^{\text{ideal}}} = \frac{h_4 - h_5}{h_4 - h_{5s}},$$

where $w_{\text{turb}}^{\text{actual}}$ is specified to be at the same pressure ratio as $w_{\text{turb}}^{\text{ideal}}$. There is a similar metric for the compressor, the compressor adiabatic efficiency:

$$\eta_{\text{comp}} = \frac{w_{\text{turb}}^{\text{ideal}}}{w_{\text{turb}}^{\text{actual}}} = \frac{h_{3s} - h_0}{h_3 - h_0},$$

again for the same pressure ratio. Note that for the turbine the ratio is the actual work delivered divided by the ideal work, whereas for the compressor the ratio is the ideal work needed divided by the actual work required. These are not thermal efficiencies, but rather measures of the degree to which the compression and expansion approach the ideal processes.

We now wish to find the net work done in the cycle and the efficiency. The net work is given either by the difference between the heat received and rejected or the work of the compressor and turbine, where the convention is that heat received is positive and heat rejected is negative and work done is positive and work absorbed is negative.

$$\text{Net work} = \begin{cases} \underbrace{q}_{\text{heat in}} - \underbrace{qR}_{\text{heat out}} = (h_4 - h_3) - (h_5 - h_0) \\ w_{\text{turb}} - w_{\text{comp}} = (h_4 - h_5) - (h_3 - h_0). \end{cases}$$

The thermal efficiency is

$$\eta_{\text{thermal}} = \frac{\text{Net work}}{\text{Heat input}}.$$

We need to calculate T_3, T_5.

From the definition of η_{comp},

$$T_3 - T_0 = \frac{(T_{3s} - t_0)}{\eta_{\text{comp}}} = T_0 \frac{(T_{3s}/T_0 - 1)}{\eta_{\text{comp}}}.$$

With

$$(T_{3s}/T_0) = \text{isentropic temperature ratio} = \left(\frac{P_{\text{exit}}}{P_{\text{inlet}}}\right)^{\frac{\gamma-1}{\gamma}} = \Pi_{\text{comp}}^{\frac{\gamma-1}{\gamma}},$$

$$T_3 = T_0 \frac{\left(\Pi_{\text{comp}}^{\frac{\gamma-1}{\gamma}} - 1\right)}{\eta_{\text{comp}}} + T_0.$$

Similarly, by the definition

$$\text{h}_{\text{turb}} = \frac{\text{actual work received}}{\text{ideal work for same}\dfrac{P_{\text{exit}}}{P_{\text{inlet}}}},$$

we can find T_5:

$$T_4 - T_5 = \eta_{\text{turb}}(T_4 - T_{5s}) = \eta_{\text{turb}} T_4 \left(1 - \frac{T_{5s}}{T_4}\right) = \eta_{\text{tutb}} T_4 \left(1 - \Pi_{\text{turb}}^{\frac{\gamma-1}{\gamma}}\right)$$

$$T_5 = T_4 - \eta_{\text{turb}} T_4 \left(1 - \Pi_{\text{turb}}^{\frac{\gamma-1}{\gamma}}\right).$$

The thermal efficiency can now be found:

$$\eta_{\text{thermal}} = 1 + \frac{Q_L}{Q_H} = 1 - \frac{T_5 - T_0}{T_4 - T_3}.$$

With

$$\Pi_{\text{comp}} = \frac{1}{\Pi_{\text{turb}}} = \Pi$$

and

$$\tau_s = \Pi^{\frac{g-1}{\gamma}} = \text{the isentropic cycle temperature ratio,}$$

$$\text{h}_{\text{termal}} = 1 - \frac{T_4\left[1 - \eta_{\text{tur b}}\left(1 - \dfrac{1}{\tau_s}\right)\right] - T_0}{T_4 - T_0\left[\dfrac{1}{\eta_{\text{comp}}}(\tau_s - 1) + 1\right]}$$

or

$$= \frac{\left[1 - \dfrac{1}{\tau_s}\right]\left[\eta_{\text{comp}}\eta_{\text{turb}}\dfrac{T_4}{T_0} - \tau_s\right]}{1 + \eta_{\text{comp}}\left[\dfrac{T_4}{T_0} - 1\right] - \tau_s}.$$

There are several non-dimensional parameters that appear in this expression for thermal efficiency. We list these in the two sections below and show their effects in accompanying.

PARAMETERS REFLECTING THE ABILITY TO DESIGN AND EXECUTE EFFICIENT COMPONENTS

η_{comp} : compressor adiabatic efficiency

η_{turb} : turbine adiabatic efficiency

In addition to efficiency, net rate of work is a quantity we need to examine,

$$\dot{W}_{net} = \dot{W}_{turbine} - \dot{W}_{compressor}.$$

Putting this in a non-dimensional form,

$$\frac{\dot{W}_{net}}{\dot{m}c_p T_0} = \underbrace{\frac{1}{\eta_{comp}}(\tau_s - 1)}_{\text{work to drive compressor}} + \underbrace{\eta_{turb}\frac{T_4}{T_0}\left(1 - \frac{1}{\tau_s}\right)}_{\text{work extracted from flow by turbine}}.$$

$$\frac{\dot{W}_{net}}{\dot{m}c_p T_0} = (t_s - 1)\left[\frac{\eta_{turb}\dfrac{T_4}{T_0}}{\tau_s} - \frac{1}{\eta_{comp}}\right].$$

Trends in net power and efficiency for parameters typical of advanced civil engines.

- For any η_{comp}, $\eta_{turb} \neq 1$ the optimum pressure ratio p_i for maximum t is not the highest that can be achieved, as it is for the ideal Brayton cycle. The ideal analysis is too idealized in this regard. The highest efficiency also occurs closer to the pressure ratio for maximum power than in the case of an ideal cycle. Choosing this as a design criterion will therefore not lead to the efficiency penalty inferred from ideal cycle analysis.
- There is a strong sensitivity to the component efficiencies. for $\sum_i p_i = 1$, the cycle efficiency is roughly two-thirds of the ideal value.
- The maximum power occurs at a value of (E) or pressure ratio p_i less than that for max

$$(E) = \sum p_i$$

(this trend is captured by ideal analysis).

- The maximum power and maximum ε_i are strongly dependent on the maximum temperature, p_i.

A scale diagram of a Brayton cycle with non-ideal compressor and turbine behaviors, in terms of temperature-entropy (*h-s*) and pressure-volume (*P-v*) coordinates.

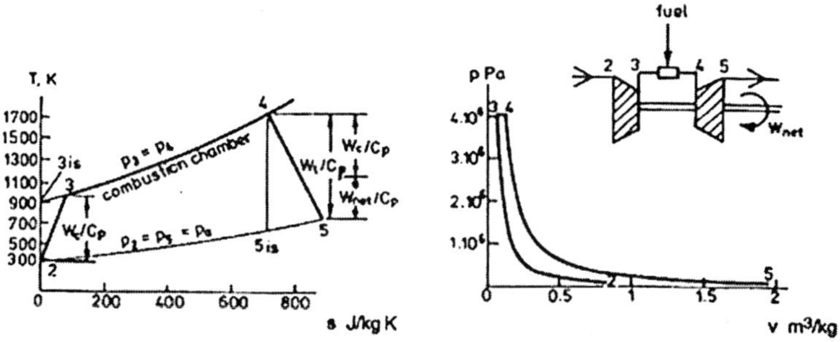

Fig. Scale Diagram of Non-ideal gas Turbine Cycle. Nomenclature Pressure ratio = 1, p_2, = 0, Compressor and Turbine Efficiencies p_3

REFERENCES

- Schaum's Outline of Thermodynamics for Engineers, 2nd edition (Schaum's Outlines) by Merle Potter and Ph.D., Craig Somerton.
- Essentials of Thermodynamics (Essentials) by Research & Education Association, Rea, and Staff of Research Education Association .
- Statistical mechanics, R. P. Feynman, W. A. Benjamin.
- A Modern Course in Statistical Physics, L. E. Reichl
- Thermodynamics by J.P. Holman.

9

Third Law of Thermodynamics

The Third Law of Thermodynamics is the lesser known of the three major thermodynamic laws. Together, these laws help form the foundations of modern science. The laws of thermodynamics are absolute physical laws - everything in the observable universe is subject to them. Like time or gravity, nothing in the universe is exempt from these laws. In its simplest form, the Third Law of Thermodynamics relates the entropy (randomness) of matter to its absolute temperature.

The Third Law of Thermodynamics refers to a state known as "absolute zero." This is the bottom point on the Kelvin temperature scale. The Kelvin scale is absolute, meaning 0° Kelvin is mathematically the lowest possible temperature in the universe. This corresponds to about -273.15° Celsius, or -459.7 Fahrenheit.

In actuality, no object or system can have a temperature of zero Kelvin, because of the Second Law of Thermodynamics. The Second Law, in part, implies that heat can never spontaneously move from a colder body to a hotter body. So, as a system approaches absolute zero, it will eventually have to draw energy from whatever systems are nearby. If it draws energy, it can never obtain absolute zero. So, this state is not physically possible, but is a mathematical limit of the universe.

In its shortest form, the Third Law of Thermodynamics says: "The entropy of a pure perfect crystal is zero (0) at zero Kelvin (0° K)." Entropy is a property of matter and energy discussed by the Second Law of Thermodynamics. The Third Law of Thermodynamics means that as the temperature of a system approaches absolute zero, its entropy approaches a constant (for pure perfect crystals, this constant is zero). A pure perfect crystal is one in which every molecule is identical, and the molecular alignment is perfectly even throughout the substance. For non-pure crystals, or those with less-than perfect alignment, there will be some energy associated with the imperfections, so the entropy cannot become zero.

The Third Law of Thermodynamics can be visualized by thinking about water. Water in gas form has molecules that can move around very freely. Water vapour has very high entropy (randomness). As the gas cools, it becomes liquid. The liquid water molecules can still move around, but not as freely. They have lost some entropy. When the water cools further, it becomes solid ice. The solid water molecules can no longer move freely, but can only vibrate within the ice crystals. The entropy is now very low. As the water is cooled more, closer and closer to absolute zero, the vibration of the molecules diminishes. If the solid water reached absolute zero, all molecular motion would stop completely. At this point, the water would have no entropy (randomness) at all.

Most of the direct use of the Third Law of Thermodynamics occurs in ultra-low temperature chemistry and physics. The applications of this law have been used to predict the response of various materials to temperature changes. These relationships have become core to many science disciplines, even though the Third Law of Thermodynamics is not used directly nearly as much as the other two.

MATHEMATICAL NOTATIONS

If we have sufficient heat capacity data (and the data on phase changes) we could write

$$S(T) = S(T = 0) = \int_0^T \frac{C_p}{T} dT.$$

(If there is a phase change between 0 K and T we would have to add the entropy of the phase change.) If C_p were constant near $T = 0$ we would have,

$$S(T) = S(T = 0) + C_P \ln\frac{T}{0},$$

which is undefined. Fortunately, experimentally $C_p \to 0$ as $T \to 0$. For nonmetals C_p is proportional to T^3 at low temperatures. For metals C_p is proportional to T^3 at low temperatures but shifts over to being proportional to T at extremely low temperatures. (The latter happens when the atomic motion "freezes out" and the heat capacity is due to the motion of the conduction electrons in the metal.)

Equation could be used to calculate absolute entropies for substances if we knew what the entropy is at absolute zero. Experimentally it appears that the entropy at absolute zero is the same for all substances. The third law of thermodynamics codifies this observation and sets

$$S(T = 0) = 0$$

for all elements and compounds in their most stable and perfect crystalline state at absolute zero and one atmosphere pressure. (All except for helium, which is a liquid at the lowest observable temperatures at one atmosphere.)

The advantage of this law is that it allows us to use experimental data to compute the absolute entropy of a substance. For example, suppose we want to calculate the absolute entropy of liquid water at 25° C. We would need to know the C_p of ice from 0 K to 273.15 K and the $\underline{C}_p$ of liquid water from 273.15 K to 298.15 K. We also need the heat of fusion of water at its normal melting point. With all of this data, which can be obtained partly from theory and partly from experiment, we find

$$S^{\circ}_{H_2O}(25°C) = 0 + \int_0^{273.15} \frac{C_P(S)}{T} dT + \frac{DH_{\text{fus}}}{273.15} + \int_{273.15}^{298.15} \frac{C_p(l)}{T} dT,$$

Some substances may undergo several phase changes.

Entropy Changes in Chemical Reactions

We can use the third law entropies to calculate entropy changes for chemical reactions. For a typical reaction,

$$a\,\text{A} + b\,\text{B} \text{ '! } c\,\text{C} + d\,\text{D}.$$

the entropy change is

$$\Delta_r S^\circ = cS_C^\circ + dS_D^\circ - aS_A^\circ - bS_B^\circ.$$

Notice two things:

- We did not define or use an entropy of formation, $Ä_f S^{\mathrm{o}}$.
- $S^{\mathrm{o}}_{\mathrm{element}}$ is not zero.

As we have said before, $Ä_r S^{\mathrm{o}}$ and $Ä_r H^{\mathrm{o}}$ are independent of each other. They can not be calculated from each other. They must be calculated from Equation and a comparable equation using heats of formation.

There is another way to calculate $Ä_r S^{\mathrm{o}}$,

$$\Delta_r S^\circ = \frac{\Delta_r H^\circ - \Delta_r G^\circ}{T}.$$

As we have seen before, the $Ä_r G^{\mathrm{o}}$ and $Ä_r H^{\mathrm{o}}$ can be calculated from free energies and heats of formation.

As $T \rightarrow 0$ K , $S \rightarrow 0$.

For the General student, the important point about the third law is that entropy is an absolute quantity which depends upon temperature. This is in contrast to ÄH for reactions which have as a reference the elemental state. Thus, when one looks up the $ÄH^{\mathrm{o}}{}_{\mathrm{f}}$ of an elements, the answer is 0. In contrast, S^{o} for an element (note difference in symbols as well) has a value for temperature above 0 K. Careful when doing calculations for $ÄS^{\mathrm{o}}$ of reactions that you do not use 0 for the S^{o} of the elements.

The entropy change with respect to temperature can be thought of a continuous summation of all the increments of heat added to the system divided by the temperature at the time of the addition. Or symbolically:

$\Delta S = \int (dq/T)dT$ which is approximately SUM of the ($Äq$ / T)s

Thus, to calculate a change in S one simply adds up the little increments of hcat added divided by temperature.

The question then is, what if the addition of these increments start with the temperature at 0 K? The answer is, that at 0K the q added is also 0. 0 divided by 0 presents a dilemma and the third law answers this by the following:

For a pure component in the most stable condition, S = 0 at T = 0 K.

This leads to the assumption needed above, that the S° s for pure components are absolute values and are not referenced against some arbitrary initial condition like the ÄH° s are. As an illustration, see the example thermodynamic table and notice that the elements do have S° s listed. Check out the following:

For the pure components (complete chemicals) the **S°s** are positive

For ions, which are not complete chemicals but only one leg of the ionic compound, there are ÄS° listed which can be either positive or negative. These ions are reference against the H^+ (understood to stand for H_3O^+) ion.

REFERENCE

- Course in Thermodynamics . Revised Printing. Volume II. (Series in Thermal and Fluids Engineering) by Joseph Kestin.
- Thermodynamics In Materials Science by Robert T. Dehoff .
- Statistical Mechanics, S-K Ma
- Schaum Engineering Thermodynamics (Schaum's Outlines) by Merle Potter.
- Thermodynamics by J.P. Holman

10

Entropy

ENTROPY CHANGE IN MIXING OF TWO IDEAL GASES

Consider an insulated rigid container of gas separated into two halves by a heat conducting partition so the temperature of the gas in each part is the same. One side contains air, the other side another gas, say argon, both regarded as ideal gases. The mass of gas in each side is such that the pressure is also the same.

The entropy of this system is the sum of the entropies of the two parts: turbine adiabatic efficiency. Suppose the partition is taken away so the gases are free to diffuse throughout the volume. For an ideal gas, the energy is not a function of volume, and, for each gas, there is no change in temperature. (The energy of the overall system is unchanged, the two gases were at the same temperature initially, so the final temperature is the same as the initial temperature.) The entropy change of each gas is thus the same as that for a reversible isothermal expansion from the initial specific volume. For a mass of ideal gas, the entropy change is $\frac{\dot{W}_{net}}{\dot{m}c_pT_0} = (T_s - 1)\left[\frac{\eta_{turb}\frac{T_4}{T_0}}{\tau_s} - \frac{1}{\eta_{comp}}\right]$. The entropy change of the system is

$$\eta turb \neq 1$$

Equation states that there is an entropy increase due to the increased volume that each gas is able to access.

Examining the mixing process on a molecular level gives

additional insight. Suppose we were able to see the gas molecules in different colors, say the air molecules as white and the argon molecules as red. After we took the partition away, we would see white molecules start to move into the red region and, similarly, red molecules start to come into the white volume. As we watched, as the gases mixed, there would be more and more of the different colour molecules in the regions that were initially all white and all red.

If we moved further away so we could no longer pick out individual molecules, we would see the growth of pink regions spreading into the initially red and white areas. In the final state, we would expect a uniform pink gas to exist throughout the volume. There might be occasional small regions which were slightly more red or slightly more white, but these fluctuations would only last for a time on the order of several molecular collisions.

In terms of the overall spatial distribution of the molecules, we would say this final state was more random, more mixed, than the initial state in which the red and white molecules were confined to specific regions. Another way to say this is in terms of "disorder;" there is more disorder in the final state than in the initial state. One view of entropy is thus that increases in entropy are connected with increases in randomness or disorder. This link can be made rigorous and is extremely useful in describing systems on a microscopic basis. While we do not have scope to examine this topic in depth, the purpose of is to make plausible the link between disorder and entropy through a statistical definition of entropy.

MICROSCOPIC AND MACROSCOPIC DESCRIPTIONS OF A SYSTEM

The microscopic description of a system is the complete description of each particle in this system. In the above example, the microscopic description of the gas would be the list of the state of each molecule: position and velocity in this problem. It would require a great deal of data for this description; there are roughly 10^{19} molecules in a cube of air one centimeter on a side at room temperature and pressure. The macroscopic description, which is in terms of a few (two!) properties is thus far more accessible and useable for engineering applications, although it is restricted to equilibrium states.

To address the description of entropy on a microscopic level, we need to state some results concerning microscopic systems. These results and the computations and arguments below are taken almost entirely from the excellent of Engineering Thermodynamics by Reynolds and Perkins.

For a given macroscopic system, there are many microscopic states. A key idea from quantum mechanics is that the states of atoms, molecules, and entire systems are discretely quantized. This means that a system of particles under certain constraints, like being in a box of a specified size, or having a fixed total energy, can exist in a finite number of allowed microscopic states. This number can be very big, but it is finite. The microstates of the system keep changing with time from one quantum state to another as molecules move and collide with one another. The probability for the system to be in a particular quantum state is defined by its quantum-state probability p_i. The set of the p_i is called the distribution of probability. The sum of the probabilities of all the allowed quantum states must be unity, hence for any time t,

$$\sum_i p_i = 1$$

When the system reaches equilibrium, the individual molecules still change from one quantum state to another. In equilibrium, however, the system state does not change with time; so the probabilities for the different quantum states are independent of time. This distribution is then called the equilibrium distribution, and the probability p_i can be viewed as the fraction of time a system spends in the ith quantum state. In what follows, we limit consideration to equilibrium states.

We can get back to macroscopic quantities from the microscopic description using the probability distribution. For instance, the macroscopic energy of the system would be the weighted average of the successive energies of the system (the energies of the quantum states); the energies are weighted by the relative time the system spends in the corresponding microstates. In terms of probabilities, the average energy, ⟨E⟩, is

$$\langle E \rangle = \sum_i p_i \varepsilon_i,$$

where ε_i is the energy of a quantum state.

The probability distribution provides information on the randomness of the equilibrium quantum states suppose the system can only exist in three states (1, 2 and 3). If the distribution probability is

$$p_1 = 1, \ p_2 = 0 \ p_3 = 0,$$

the system is in quantum state 1 and there is no randomness. If we were asked what quantum state the system is in, we would be able to say it is always in state 1. If the distribution were

$$p_1 = 0.3, \ p_2 = 0.2, \ p_3 = 0.5,$$

or

$$p_1 = 0.5, \ p_2 = 0.2, \ p_3 = 0.3,$$

the randomness would not be zero and would be equal in both cases. We would be more uncertain about the instantaneous quantum state than in the first situation.

Maximum randomness corresponds to the case where the three states are equally probable:

$$p_1 = 1/3, \ p_2 = 1/3, \ p_3 = 1/3,$$

In this case, we can only guess the instantaneous state with 33% probability.

STATISTICAL DEFINITION OF ENTROPY

The list of the p_i is a precise description of the randomness in the system, but the number of quantum states in almost any industrial system is so high this list is not useable. We thus look for a single quantity, which is a function of the p_i, that gives an appropriate measure of the randomness of a system. As shown below, the entropy provides this measure.

There are several attributes that the desired function should have. The first is that the average of the function over all of the microstates should have an extensive behaviour. In other words the microscopic description of the entropy of a system C, composed of parts A and B should be given by

$$S_C = S_A + S_B$$

Second is that entropy should increase with randomness and should be largest for a given energy when all the quantum states are equiprobable.

The average of the function over all the microstates is defined by

$$S = \langle f \rangle = \sum_i p_i f(p_i),$$

where the function $f(p_i)$ is to be found. Suppose that system A has $n$microstates and system Bhas $m$microstates. The entropies of systems A, B, and C, are defined by

$$S_A = \sum_{i=1}^{n} p_i f(p_i)$$

$$S_A = \sum_{j=1}^{m} p_i f(p_j)$$

$$S_C = \sum_{i=1}^{n}\sum_{j=1}^{m} p_i f(p_j) = \sum_{i=1}^{n} p_i f(p_i) \sum_{j=1}^{m} p_i f(p_j)$$

In Equations, the term p_{ij} means the probability of a microstate in which system A is in state i and system B is in state j. For Equation to hold given the expressions in Equations ,

$$S_C = \sum_{i=1}^{n} p_i f(p_i) \sum_{j=1}^{m} p_i f(p_j)$$

$$= \sum_{i=1}^{n} p_i f(p_i) + \sum_{j=1}^{m} p_j f(p_j) = S_A + S_B.$$

The function f must be such that this is true regardless of the values of the probabilities p_i and p_j. This will occur if $f() = \ln ()$ because ln $(a \,.\, b) = \ln (a) + \ln (b)$.

To verify this, make this substitution in the expression for S_c in the first part of Equation (assume the probabilities p_i and p_j are independent, such that $p_{ij} = p_i p_j$, and split the log term):

$$S_C = \sum_{i=1}^{n}\sum_{j=1}^{m} p_i p_j \ln(p_i) + \sum_{i=1}^{n}\sum_{j=1}^{m} p_i p_j \ln(p_j).$$

Rearranging the sums, becomes

$$S_C = \sum_{i=1}^{n}\left\{ p_i \ln(p_i)\left[\sum_{j=1}^{m} p_j\right]\right\} + \sum_{j=1}^{m}\left\{ p_j \ln(p_i)\left[\sum_{j=1}^{m} p_i\right]\right\}.$$

Because

$$\sum_{i=1}^{n} p_j = \sum_{j=1}^{m} p_j = 1,$$

the square brackets in the right hand side of Equation can be set equal to unity, with the result written as

$$S_C = \sum_{i=1}^{n} p_j \ln(p_i) + \sum_{j=1}^{m} p_j \ln(p_j).$$

This reveals the top line of Equation to be the same as the bottom line, for any p_i, p_j, n, m, provided that $f()$ is a logarithmic function. Reynolds and Perkins show that the most general $f(p_i)$ is $f = C \ln(p_i)$, where is an arbitrary constant. Because the p_i are less than unity, the constant is chosen to be negative to make the entropy positive.

Based on the above, a statistical definition of entropy can be given as:

$$S = -k\sum_{i} p_i \ln(p_i).$$

The constant k is known as the Boltzmann constant,

$$k = 1.380 \times 10^{-23}\frac{J}{K}.$$

The value of is (another wonderful result!) given by

$$k - \frac{R}{N_{\text{Avogadro}}},$$

where R is the universal gas constant, 8.3143 J/(mol-K) and N_{Avogadro} is Avogadro's number, 6.02×10^{23} molecules per mol. Sometimes k is called the gas constant per molecule. With this value for k, the statistical definition of entropy is identical with the macroscopic definition of entropy.

CONNECTION BETWEEN THE STATISTICAL DEFINITION OF ENTROPY AND RANDOMNESS

We need now to examine the behaviour of the statistical definition of entropy as regards randomness. Because a uniform probability distribution reflects the largest randomness, a system with allowed states will have the greatest entropy when each state is equally likely. In this situation, the probabilities become

$$p_i = p = \frac{1}{\Omega},$$

where $\Omega$is the total number of microstates. The entropy is thus

$$S = -k\sum_{i=1}^{\Omega}\frac{1}{\Omega}\ln\left(\frac{1}{\Omega}\right) = -k\Omega\left[\frac{1}{\Omega}\ln\left(\frac{1}{\Omega}\right)\right] = -k\ln\left(\frac{1}{\Omega}\right)$$
$$= k\ln\Omega.$$

Equation states that the larger the number of possible states the larger the entropy. The behaviour of the entropy stated in Equation can be summarized as follows:

- S is maximum when Ω is maximum, which means many permitted quantum states, hence much randomness,
- S is minimum when Ω is minimum. In particular, for $\Omega = 1$, there is no randomness and $S = 0$.

These trends are in accord with our qualitative ideas concerning randomness. Equation is carved on Boltzmann's tombstone.

We can also examine the additive property of entropy with respect to probabilities. If we have two systems, A and B, which are viewed as a combined system, C, the quantum states for the combined system are the combinations of the quantum states from A and B. The quantum state where A is in its state x and B is in its state would have a probability $pAx \cdot pBy$because the two probabilities are independent. The number of probabilities for the combined system, Ω_C, is thus defined by $\Omega_C = \Omega_A.\Omega_B$. The entropy of the combined system is

$$S_C = k\ln(\Omega_A\Omega_B) = k\ln\Omega_A + k\ln\Omega_B = S_A + S_B$$

Equation is sometimes taken as the basic definition of entropy, but it should be remembered that it is only appropriate when each quantum state is equally likely. Equation is more general and applies equally for equilibrium and non-equilibrium situations. A simple numerical example shows trends in entropy changes and randomness for a system which can exist in three states. Consider the five probability distributions

(i) $p_1 = 1.0, p_2 = 0, p_3 = 0;$

$$S = -k(1 \ln(1) + 0 \ln(0) + 0 \ln(0)) = 0$$

(ii) $p_1 = 0.8\ p_2 = 0.2, p_3 = 0;$

$$S = -k(0.8 \ln(0.8) + 0.2 \ln(0.2) + 0 \ln(0)) = 0.5k$$

(iii) $p_1 = 0.8\ p_2 = 0.1, p_3 = 0.1;$

$$S = -k(0.8 \ln(0.8) + 0.1 \ln(0.1) + 0.1 \ln(0.1)) = 0.6k$$

(iv) $p_1 = 0.5\ p_2 = 0.3, p_3 = 0.2;$

$$S = -k(0.5 \ln(0.5) + 0.3 \ln(0.3) + 0.2 \ln(0.2)) = 1.0k$$

$$S = -3k\left[\frac{1}{3}\ln\left(\frac{1}{3}\right)\right]$$

The first distribution has no randomness. For the second, we know that state 3 is never found. Distributions (iii) and (iv) have progressively greater uncertainty about the distribution of states and thus higher randomness. Distribution (v) has the greatest randomness and uncertainty and also the largest entropy.

REFERENCES

- Heat and Thermodynamics, M. W. Zemansky
- Thermodynamics (and Introduction to Thermostatistics), H.B. Callen
- Schaum's Outline of Thermodynamics for Engineers, 2nd edition (Schaum's Outlines) by Merle Potter and Ph.D., Craig Somerton.
- Introduction to Metallurgical Thermodynamics by David R. Gaskell.
- Equilibrium Thermodynamics, C. J. Adkins.

11

Enthalpy Generating Heat

There are a wide variety of fuels used for aerospace power and propulsion. A primary one is jet fuel (octane, essentially kerosene) which has the chemical formula C_8H_{18}. Other fuels we consider are hydrogen (H_2) and methane (CH_4).

The chemical process in which a fuel, methane, is burned consists of (on a very basic level -— there are many intermediate reactions that need to be accounted for when computations of the combustion process are carried out):

$$\underbrace{CH_4 + 2O_2}_{\text{(Reactants)}} \longrightarrow \underbrace{CO_2 + 2H_2O}_{\text{(Products)}}$$

The reactions we describe are carried out in air, which can be approximated as 21% O_2 and 79% N_2. This composition is referred to as "theoretical air." There are other components of air (Argon, which is roughly 1%), but the results given using the theoretical air approximation are more than adequate for our purposes. With this definition, for each mole of O_2, 3.76 (or 79/21) moles of N_2 are involved:

$$CH_4 + 2O_2 + 2(3.76)N_2 \longrightarrow CO_2 + 2H_2O + 7.52N_2.$$

Even if the nitrogen is not part of the combustion process, it leaves the combustion chamber at the same temperature as the other products, and this change in state (change in enthalpy) needs to be accounted for in the steady flow energy equation. At the high temperatures achieved in internal combustion engines (aircraft and automobile) reaction does occur between the nitrogen and oxygen, which gives rise to oxides of nitrogen, although we will not consider these reactions. The condition at which the mixture of fuel and air is such that both completely participate in the reaction is called stoichiometric. In gas turbines, excess air is often used so that the temperatures of the gas exiting the combustor is kept to within desired limits.

Fuel-Air Ratio

The reaction for aeroengine fuel at stoichiometric conditions is

$$C_8H_{18} + 12.5\ O_2 + 12.5\ (3.76)N_2 \longrightarrow 8CO_2 + 9H_2O + 47.0\ N_2.$$

On a molar basis, the ratio of fuel to air is $[1/(12.5 + 47.0)]=1/59.5 = 0.0167$.

To find the ratio on a mass flow basis, which is the way in which the aeroengine industry, we need to "weight" the molar proportions by the molecular weight of the components. The fuel molecular weight is 114 g/mol, the oxygen molecular weight is 32 g/mol and the nitrogen molecular weight is (approximately) 28 g/mol. The fuel/air ratio on a mass flow basis is thus

Fuel-air ratio:

$$= \frac{1\ \text{mol} \times 114\ \text{g/mol}}{12.5\ \text{mol} \times 32\ \text{g/mol} + 12.5 \times 3.76\ \text{mol} \times 28\ \text{g/mol}} = 0.0664$$

If we used the actual constituents of air we would get 0.0667, a value about 0.5% different.

Enthalpy of formation

The systems we have worked with until now have been of fixed chemical composition. Because of this, we could use thermodynamic properties relative to an arbitrary base, since all comparisons could be made with respect to the chosen base. The specific energy $u_f(0.01°\text{C}) = 0.0$ for steam. If there are no changes in composition, and only changes in properties of given substances, this is adequate. If there are changes in composition, however, we need to have a reference state so there is consistency for different substances.

The convention used is that the reference state is a temperature of 25°C (298 K) and a pressure of 0.1 MPa. (These are roughly room conditions.) At these reference conditions, the enthalpy of the elements (oxygen, hydrogen, nitrogen, carbon, etc.) is taken as zero.

The results of a combustion process can be diagrammed. The reactants enter at standard conditions; the combustion (reaction) takes place in the volume indicated. Downstream of the reaction zone there is an appropriate amount of heat transfer with the surroundings so that the products leave at the standard conditions. For the reaction of carbon and oxygen to produce CO_2, the heat that has to be extracted is $Q_{CV} = -393,522$ kJ/mole; this is heat that comes out of the control volume.

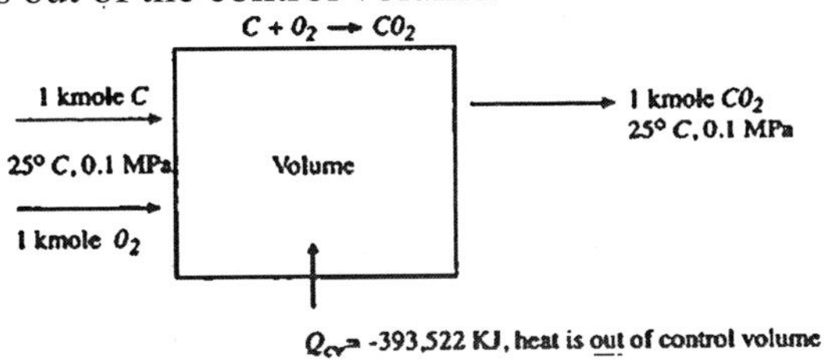

Fig. Constant Pressure Combustion

There is no shaft work done in the control volume and the first law for the control volume (SFEE) reduces to:

mass flow of enthalpy in + rate of heat addition = mass flow of enthalpy out.

We can write this statement in the form

$$\sum_R \dot{m}_i h_i + \dot{Q}_{CV} = \sum_P \dot{m}_c h_c.$$

In Equation the subscripts "R" and "P" on the summations refer to the reactants (R) and products (P) respectively. The subscripts on the mass flow rates and enthalpies refer to all of the components at inlet and at exit.

The relation in terms of mass flows can be written in molar form, which is often more convenient for reacting flow problems, by using the molecular weight, M_i, to define the molar mass flow rate, $\dot{n}_i$, and molar enthalpy, $\bar{h}_i$, for any individual ith (or eth) component as

$\dot{n}_i = \dot{m}_i$/Mi; mass flow rate in terms of kmoles/sec

$\bar{h}_i = M_i h_i$; enthaply per kmole.

The SFEE is, in these terms,

$$\sum_R \dot{n}_i \bar{h}_i + \dot{Q}_{CV} = \sum_P \dot{n}_i \bar{h}_c.$$

The statements that have been made do not necessarily need to be viewed in the context of flow processes. Suppose we have one unit of C and one unit of Q_2 at the initial conditions and we carry out a constant pressure reaction at ambient pressure, P_{amb}. If so,

$$U_{\text{final}} - U_{\text{initial}} = \text{Q} - \text{W}$$
$$= Q_{CV} - P_{\text{amb}} (V_{\text{final}} - V_{\text{initial}}),$$

since $P_i = P_f = P_{amb}$. Combining terms,

$$U_{\text{final}} + P_{\text{final}} V_{\text{final}} - (U_{\text{intial}} + P_{\text{intial}} V_{\text{initial}}) = Q_{CV},$$

or

$$H_{\text{final}} - H_{\text{intial}} = Q_{CV}.$$

In terms of the numbers of moles and the specific enthalpy this is

$$\sum_R \dot{n}_i \bar{h}_i + Q_{CV} = \sum_P \dot{n}_i \bar{h}_c.$$

The enthalpy of CO_2, at 25°C and 0.1 MPa, with reference to a base where the enthalpy of the elements is zero, is called the enthalpy of formation and denoted by $\bar{h}_f^{\circ}$. Values of the heat of formation for a number of substances are given in Table in SB&VW.

The enthalpies of the reactants and products for the formation of CO_2 are:

$$\bar{h}_{O_2} = \bar{h}_{C} = 0,$$

For one kmoke: $Q_{CV} = \sum_P n_c \bar{h}_c = H_P = \left(\bar{h}_f^{\circ}\right) CO_2 = -\ 393,\ 522$ kJ/kmole.

The enthalpy of CO_2 in any other state (T, P) is given by

$$\bar{h}_{T,P} = \left(\bar{h}_f^{\circ}\right)_{398\text{ K},\ 0.1\text{ MPA}} = \left(\Delta\bar{h}\right)_{298\text{ K},\ 0.1\text{ MP}a \to T,P.}$$

These descriptions can be applied to any compound. For elements or compounds that exist in more than one state at the reference conditions (carbon exists as diamond and as graphite), we also need to specify the state.

Note that there is a minus sign for the heat of formation. The heat transfer is out of the control volume and is thus negative by our convention. This means that $\overline{h}_{\text{elements}} > \overline{h}_{CO_2} = -\ 393{,}522$ kJ/kmole.

First Law Analysis of Reacting Systems

The form of the first law for the control volume is (there is no shaft work):

$$\sum_R \dot{n}_i \overline{h}_i + Q_{CV} = \sum_P \dot{n}_c \overline{h}_c.$$

This is given in terms of the moles of the different constituents, and it reduces to the more familiar form for a single fluid (say air) with no reactions occurring. We need to specify one parameter as the basis of the solution; 1 kmole of fuel, 1 kmole of air, 1 kmole total, etc. We use 1 kmole of fuel as the basic unit and examine the burning of hydrogen.

$$2H_2 + O_2 \longrightarrow 2H_2O$$

The reactants and the products are both taken to be at 0.1 MPa and 25°C, so the inlet and exit P and T are specified. The control volume is the combustion chamber. There is no shaft work done and the SFEE is in the form of Equation. The enthalpy of the entering gas is zero for both the hydrogen and the oxygen (elements have enthalpies defined as zero at the reference state). If the exit products are in the gaseous state, the exit enthalpy is therefore related to the enthalpy of formation of the product by:

$$\dot{n}_c, H_2O\overline{h}_c, H_2O = \dot{n}_c, H_2O\left(\overline{h}_f^{\circ}\right)H_2O(g)$$

$= 2 \times (-241{,}827)$ kJ $= -\ 483{,}654$ kJ; gaseous state at exit.

If the water is in a liquid state at the exit of the process:

$$\dot{n}_c, H_2O\overline{h}_c, H_2O = \dot{n}_c, H_2O\left(\overline{h}_f^{\circ}\right)H_2O(1)$$

$$= 2 \times (-285{,}783) \text{ kJ} = -\ 571{,}676 \text{ kJ}.$$

There is more heat given up if the products emerge as liquid.

The difference between the two values is the enthalpy needed to turn the liquid into gas at: 25°C: $h_{fg} = 2442$ kJ/kg.

A more complex example is provided by the burning of methane (natural gas) in oxygen, producing

$$CH_4 + 2O_2 \rightarrow CO2 + 2H_2O(l).$$

The components in this reaction equation are three ideal gases (methane, oxygen, and CO_2) and liquid water. We again specify that the inlet and exit states are at the reference conditions so that:

$$\sum_R \dot{n}_i \bar{h}_i = (\bar{h}_f^{\circ})_{CH_4} = -74{,}873 \text{ kJ}$$

$$\sum_P n_c \bar{h}_c = (\bar{h}_f^{\circ})_{CO_2} + 2(\bar{h}_f^{\circ})H_2O(1)$$

$$= -393.522 + 2(-285{,}838) = -965{,}198 \text{ kJ}$$

$$Q_{CV} = -965{,}198 - (-74{,}873) = -890{,}325 \text{ kJ}$$

Suppose the substances which comprise the reactants and the products are not at 25°C and 0.1 MPa. If so, the expression that connects the reactants and products is

$$Q_{CV} + \sum_R n_i \left(\bar{h}_f^{\circ} + \underbrace{\Delta\bar{h}}_{\substack{\text{Between } T_i, P_i \text{ and} \\ \text{reference conditions}}} \right) \sum_P n_c \left(\bar{h}_f^{\circ} + \underbrace{\Delta\bar{h}}_{\substack{\text{Between } T_i, P_i \text{ and} \\ \text{reference conditions}}} \right).$$

Equation shows that we must compute the enthalpy difference $\Delta\bar{h}$ between the reference conditions and the given state if the inlet or exit conditions are not the reference pressure and temperature.

There are different levels of approximation for the computation:

- Assume the specific heat is constant over the range at some average value,
- Use the polynomial expressions in the integral,
- Use tabulated values.

The first is the simplest and the crudest. Combustion processes often involve changes of a thousand degrees or more and, the specific heat for some gases can change by a factor of two or more over this range, although the changes for air are more modest.

This means that, depending on the accuracy desired, one may need to consider the temperature dependence of the specific heat in computing $\Delta\bar{h}$.

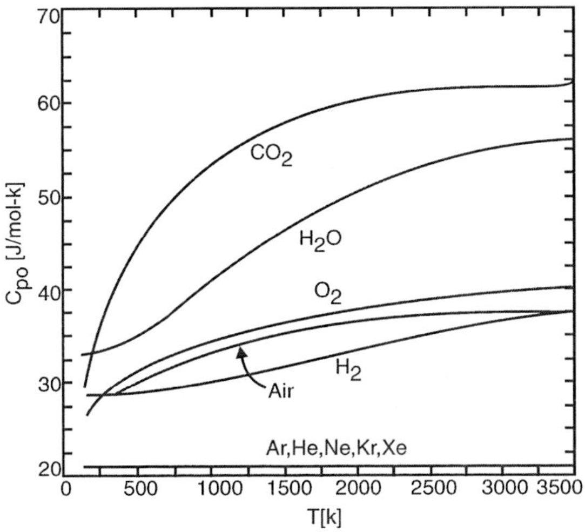

Fig. Specific heat as a function of temperature [from SB&VW]

Adiabatic Flame Temperature

For a combustion process that takes place adiabatically with no shaft work, the temperature of the products is referred to as the adiabatic flame temperature.

This is the maximum temperature that can be achieved for given reactants. Heat transfer, incomplete combustion, and dissociation all result in lower temperature.

The maximum adiabatic flame temperature for a given fuel and oxidizer combination occurs with a stoichiometric mixture (correct proportions such that all fuel and all oxidizer are consumed).

The amount of excess air can be tailored as part of the design to control the adiabatic flame temperature. The considerable distance between present temperatures in a gas turbine engine and the maximum adiabatic flame temperature at stoichiometric conditions, based on a compressor exit temperature of (922 K).

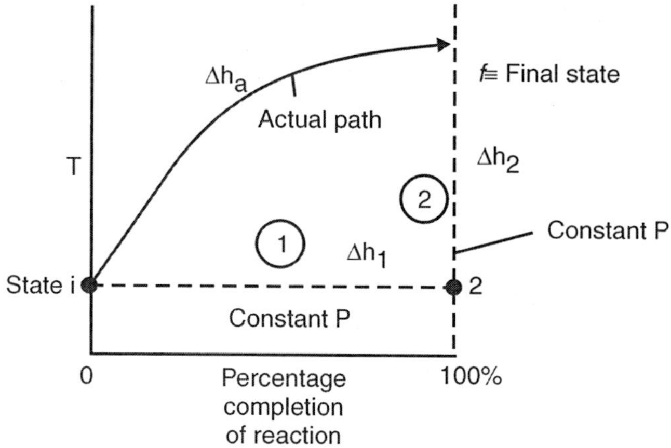

Fig. Schematic of adiabatic flame temperature

An initial view of the concept of adiabatic flame temperature is provided by examining two reacting gases, at a given pressure, and asking what the end temperature is. The process is shown schematically, where temperature is plotted versus the percentage completion of the reaction. The initial state is i and the final state is f, with the final state at a higher temperature than the initial state. The solid line in the figure shows a representation of the "actual" process.

To see how we would arrive at the final completion state the dashed lines break the state of reaction change into two parts. Process (1) is reaction at constant T and P. To carry out such a process, we would need to extract heat. Suppose the total amount of heat extracted per unit mass is q_1. The relation between the enthalpy changes in Process (1) is

$$h_2 - h_i - q_1 = (\bar{h}_f^{\circ})_{\text{unit mass}},$$

where q_1 is the "heat of reaction."

For Process (2), we put this amount back into the products to raise their temperature to the final level. For this process,

$$h_f - h_2 = q_1,$$

or, if we can approximate the specific heat as constant (using some appropriate average value)

$$c_{\text{p, avg}}\,(T_1 - T_2) = q_1.$$

For the overall process there is no work done and no heat exchanged so that the difference in enthalpy between initial and final states is zero:

$$\Delta h_1 - \Delta h_2 = \Delta h_{\text{adiabatic}} = 0.$$

The temperature change during this second process is therefore given by (approximately)

$$(T_f - T_2) = \frac{q_1}{c_{\text{p, avg}}} = \frac{|(h_f^{\circ})_{\text{unit mass}}|}{c_{\text{p, avg}}}.$$

The value of the adiabatic flame temperature given in Equation is for 100% completion of the reaction. In reality, as the temperature increases, the tendency is for the degree of reaction to be less than 100%. For the combustion of hydrogen and oxygen, at high temperatures the combustion product (water) dissociates back into the simpler elemental reactants. The degree of reaction is thus itself a function of temperature that needs to be computed.

We used this idea in discussing the stoichiometric ramjet, when we said that the maximum temperature was independent of flight Mach number and hence of inlet stagnation temperature. It is also to be emphasized that the idea of a constant (average) specific heat, $c_{\text{p, avg}}$, is for illustration and not inherently part of the definition of adiabatic flame temperature.

An example computation of adiabatic flame temperature is furnished by the combustion of liquid octane at 25°C with 400% theoretical air. The reaction is

$$C_8H_{18}(l) + 12.5O_2 + 12.5(3.76N_2) + 3[12.5O2 + 12.5(3.76N_2)]$$
$$\rightarrow 8CO_2 + 9H_2O\ (g) + 37.5O_2 + 188N_2.$$

For an adiabatic process

$$\sum_R n_i(h_f^{\circ} + \Delta\bar{h})_i = \sum_P n_c(h_f^{\circ} + \underbrace{\Delta\bar{h}}_{\substack{\text{At adiabatic}\\ \text{flame temperature}}})_c.$$

We can again think of the general process in steps:

- Bring reactants to 25°C [the term $(\Delta\bar{h})$] from the initial temperature, using whatever heat transfer, q_1, is needed. In this example we do not need step (i) because we are already at the reference temperature.

- Reaction at 25°C [the term $(h_f^\circ)_{\text{reactants}\to\text{products}}$]. There will be some heat transfer in this step, q_a, out of the combustor.
- Put back heat $q_a + q_b$ into the products of combustion. The resulting temperature is the adiabatic flame temperature.

In the present case Equation is, explicitly:

$(h_f^\circ)C_8H_{18}(\text{l}) = 8(h_f^\circ)CO_2 + 9(h_f^\circ)H_2O + \{\Delta\bar{h}\,CO_2 + 9\Delta\bar{h}\,H_2\text{O} + 37.5\,\Delta\bar{h}\text{O}_2 + 188\Delta\bar{h}\,N_2\}$.

We can examine the terms in the SFEE separately, starting with the heat of formation terms, and keeping track of units:

$$h_f^\circ: 8(\bar{h}_f^\circ)\ CO_2 + 9(\bar{h}_f^\circ)\ H_2O - (\bar{h}_f^\circ)\ C_8H_{18}\ (l)$$

= 8 kmole (–393,522 kJ/kmole) + 9 kmole (–241,827 3 kJ/kmole)

–1 kmole (–249,952 kJ/kmole)

= – 5.075 · 10[6] kJ.

The exit state at the adiabatic flame temperature is specified by:

$$\sum_P n_c \Delta\bar{h}_c = 5.075 \times 10^6 \text{ kJ}.$$

We find the adiabatic flame temperature in three ways:

- an approximate solution using an average value of c_p,
- a more accurate one using the tabulated evolution of c_p with temperature,

Approximate Solution Using "Average" Values of Specific Heat

We can use the values at 500 K as representative. These are:

Gas	c_p (kJ/mole)
CO_2	45
H_2O	35
O_2	30
N_2	30

Using $\Delta\bar{h} = \bar{c}_{p,\,avg}\Delta T$,

$$\sum_P n_c \Delta \bar{h}_c = \Delta T\{8\,(c_p)\,CO_2 + 9\,(c_p)\,H_2O + 37.5\,(c_p)\,O_2 + 188\,(c_p)N_2\},$$

where $\Delta T = T_{\text{final}} - 25°\text{C} = T_{\text{final}} - 298$ K.

$$\sum_P n_c \Delta \bar{h} = 744 \Delta T \text{ kJ/K},$$

and using the exit state calculated above, find that

$$\Delta \text{T} = 682 \text{ K} \Rightarrow T_{\text{final}} = 980 \text{ K}.$$

Solution for adiabatic flame temperature using evolutions of specific heats with temperature

Tables give the following evolutions of specific heats with temperature:

Gas Evolution of $c_p \sqrt{R}$ with T (kJ/kmol)

CO_2 $2.401 + (8.735 \times 10^{-3})T - (6.607 \times 10^{-6})T^2 + (2.002 \times 10^{-9})T^3$

H_2O, $4.070 - (1.108 \times 10^{-3})T + (4.152 \times 10^{-6})T^2 - (2.964 \times 10^{-9})T^3 + (0.156 \times 10^{-12})T^4$

O_2, $3.626 - (1.878 \times 10^{-3})T + (7.055 \times 10^{-6})T^2 - (6.764 \times 10^{-9})T^3 + (2.156 \times 10^{-12})T^4$

N_2, $3.675 - (1.208 \times 10^{-3})T + (2.324 \times 10^{-6})T^2\ 3 - (0.632 \times 10^{-9})T^3 + (0.226 \times 10^{-12})T^4$

Using

$$\Delta h = \int_{298}^{T_f \text{ in K}} c_p(T)dT$$

and the same equation as above, we obtain

$$T_f = 899 \text{ K}.$$

Solution for adiabatic flame temperature using tabulated values for gas enthalpy

$$\Delta \bar{h}_{CO_2} \quad \Delta \bar{h}_{H_2O} \quad \Delta \bar{h}_{O_2} \quad \Delta \bar{h}_{N_2}$$

T = 28,041–21,924–19,246–18,221 kJ/kmole

T = 33,405–25,978–22,707–21,460 kJ/kmole

Plugging in the numbers shows the answer is between these two conditions. Linearly interpolating gives a value of $T_{final} = 962$ K.

REFERENCES

- Introductory Statistical Mechanics, R. Bowley and M. Sánchez
- Thermodynamics In Materials Science by Robert T. Dehoff .
- The Cluster Expansion, W. J. Mullin
- The Language of Science by Sidney B. Cahn.
- Course in Thermodynamics . Revised Printing.Volume II. (Series in Thermal and Fluids Engineering) by Joseph Kestin

12

Isolated Paramagnets

MICROCANONICAL ENSEMBLE OF ISOLATED PARAMAGNET

The mobile electrons in a semiconductor, carrying electrical charge under any applied voltage, are the electrons occupying the energy states of conduction band ($E > E_c$). The counting statistics of conduction band electrons is based on number of electrons = (number of states) * (probability of occupancy by an electron).

The number of electron energy states per unit volume over the energy interval, E ~ E+dE, is defined as

Number of states = g(E)dE.

Here g(E) is called the density of states (DOS).

Then, an energy state at energy E has a definite probability to be occupied by an electron given by

Probability of occupancy = f(E).

This probability is the main conclusion of the Fermi-Dirac statistics and f(E) is called the Fermi distribution function.

Therefore, the number of electrons over the energy interval, E ~ E+dE, is found as number of electrons overE ~ E + dE = (number of states) * (probability of occupancy)

= g(E)dE * f(E)

Total number of conduction band electrons is found by integrating this:

n = g(E)f(E)dE

over E = Ec ~ +infinity.

For holes in the valence band (i.e., the unoccupied states in VB),

number of holes over E ~ E+dE = (number of states) * (probability of vacancy)

= g(E)dE * [1-f(E)]

And the total number of holes,

p = g(E)[1-f(E)]dE

over E = -infinity ~ Ev.

ARRANGEMENT OF SUBATOMIC PARTICLES IN AN ATOM

Rutherford's Scattering Experiment:

- *Experimental set-up:* Positively charged particles (alpha) were focused at a thin (0.00004 cm) sheet of gold foil.
- *Expectation:* The alpha particles would scatter as they were deflected by the gold atoms, producing a pattern similar to spray from a nozzle.
- *Observation 1:* Most alpha particles passed straight through the gold foil.
  Conclusion: Volume taken up by atoms is mostly empty space.
- *Observation 2:* A few alpha particles bounced back toward the source.
- *Conclusion:* Particles must have hit something that was tiny, dense and electrically charged.
  Rutherford named the tiny, dense, positively charged region at the centre of the atom the nucleus.
- Protons and neutrons are found in the nucleus. Electrons occupy a large region of space around the nucleus (electron cloud) and are in motion.
- Atomic number: number of protons. Each element has a different atomic number.
- Elements in the periodic table are listed according to increasing atomic number.
- Mass number: total number of protons and neutrons in nucleus of an atom

ARRANGEMENT OF ELECTRONS IN ATOMS

An element's properties are determined largely by the number of electrons in its atoms and how these electrons are arranged.

The Bohr Model

- Electrons occupy different energy levels in the space surrounding an atomic nucleus.
- Electrons under normal circumstances maintain the lowest possible energy level.

(Ground state or unexcited state)

- Electrons can absorb energy and "jump" to a higher energy level.

(Excited state)

- When returning to the ground state, the electron gives off energy as a particular blend of colours.
- Separating this blend of colours produces an element's line spectrum (i.e. fingerprint).
- The energy difference between levels is specific to a given element, resulting in a specific colour emission:

Element	Colour
K	lavender
Ca	orange-red
Na	yellow
Ba	green

- The significance of line spectra is that energy can be given off or absorbed only in definite amounts (quanta).
- These specific amounts of energy relate to differences between energy levels:
- Electrons occupy the lowest energy level available until it is full.
- The reactivity of elements is due to the number of electrons in the outermost energy level (valence electrons).

- The construction of the periodic table indicates the number of valence electrons for each element.

ELECTROMAGNETIC RADIATION

All types of radiant energy can be described as waves, with a specific wavelength (λ) and frequency (ν) .

- The speed at which the wave is travelling is equal to the wavelength times the frequency.

$\lambda v = c$

- Light corresponds to a portion of the electromagnetic spectrum, and c represents the speed of light, 3.00 10^8 m/s.
- When passed through a prism, white light can be separated into a continuous spectrum of colours: ROYGBIV.
- Frequency and wavelength change with the progression of colours. Red ($\lambda \sim 7\ 10^{-7}$ m, $v \sim 4.3\text{x}\ 10^{14}$ Hertz), Violet ($\lambda \sim 4\text{x}10^{-7}$ m, $v \sim 7.5\text{x}10^{14}$ Hertz)
- 1 Hertz = 1/s = 1 s^{-1}
- 1 Angstrom (A) = $1\text{X}10^{-10}$ m
- Light can also be described as composed of particles called photons, with each photon having a particular amount (quantum) of energy.
- The energy of a photon of light is given by Planck's equation:

E = hv or E = hc/l

- Planck's constant = h = 6.6262 10^{-34} J•s
- Energy is directly proportional to frequency and inversely proportional to wavelength.

The red colour in many fireworks is due to the emission of light from strontium salts.

Calculate the frequency of a photon of light of wavelength 6.50 103 A. Also calculate the energy in kJ for one mole of these photons.

$\lambda v = c$ $\qquad\qquad$ $E = hv$

In the 19th century J. R. Rydberg developed an equation based on observations showing a relationship between the wavelengths of the lines in the hydrogen line spectrum.

$1/\lambda = R\,(1/n_1^2 - 1/n_2^2)$

R = Rydberg constant = $1.097\ 10^7\ m^{-1}$

n_1 and n_2 are integers such that $n_2 > n_1$

Niels Bohr further explained this relationship by suggesting that n_1 and n_2 represent energy levels where electrons can exist in the hydrogen atom.

- He assumed that these levels corresponded to quantized amounts of energy, so that electrons could only exist at these levels and must absorb or emit a specific amount of energy to move to a different level.
- Bohr described these energy levels as circular orbits around the nucleus.

Calculate the wavelength and energy of a photon of light needed to promote an electron from the 1st energy level to the 4th energy level in a hydrogen atom.

Rydberg's equation and Bohr's model are able to explain the behaviour of an electron in a one electron species (H atom, He^+ ion, etc.), but do not explain more complex species.

- A more accurate explanation of how electrons are arranged and what types of transitions are possible is needed.
- Electrons in atoms behave more like waves than particles.

Heisenberg Uncertainty Principle

It is impossible to determine accurately both the momentum and the position of an electron simultaneously.

Instead, we can describe the probability of finding an electron within a specific region, using quantum numbers.

- Atomic orbital: a region of space where there is a high probability of finding an electron.
- Quantum numbers are used to describe electrons in possible atomic orbitals.

Principal quantum number (n) ... describes the main energy level in which the electron is found

- Possible values: n = 1, 2, 3, 4, etc. with each level existing further out from the nucleus

Subsidiary quantum number (l) ... describes the shape of atomic orbital; these shapes are referred to as sublevels

- Possible values within a main energy level: $\lambda = 0, 1, 2, \ldots (n - 1)$
- Energy level n=1 has 1 sublevel (λ =0)

  Energy level n = 2 has 2 sublevels (λ =0, 1)

  Energy level n = 3 has 3 sublevels (λ =0, 1, 2)

  Energy level n = 4 has 4 sublevels (λ =0, 1, 2, 3)
- Sublevel λ = 0 is a "s" sublevel (spherical)

  Sublevel λ = 1 is a "p" sublevel (hour glass)

  Sublevel λ = 2 is a "d" sublevel

  Sublevel λ = 3 is a "f" sublevel
- Relative arrangement of s sublevels:

Magnetic quantum number ($m\lambda$) ...describes the spatial orientation of an atomic orbital

- Possible values within a sublevel:

$m\lambda = -\lambda, \ldots, 0, \ldots, \lambda$

- s sublevels have 1 possible orientation ($m\lambda$ = 0)

  p sublevels have 3 possible orientations ($m\lambda$ = -1, 0, 1)

  d sublevels have 5 possible orientations ($m\lambda$ = -2, -1, 0, 1, 2)

  f sublevels have 7 possible orientations ($m\lambda$ = -3, -2, -1, 0, 1, 2, 3)

Spin quantum number (m_s) ... describes the spin of an electron and the orientation of the magnetic field produced by the spin.

- Possible values: m_s = -½, ½

  Atomic orbitals can accommodate a maximum of two electrons, each with opposite spin.

- Electrons in the same orbital with opposite spins are spin-paired, often called paired.

How many orbitals exist and how many electrons can fit in…
Level 1?
Level 2?
Level 3?
Level 4?
Energy

Level	Sublevel	Orbitals	Electrons
n = 1	**s**	**1**	**2**
n = 2	s,p	1+3=4	8
n = 3	s, p, d	1+3+5=9	18
n = 4	s, p, d, f	1+3+5+7=16	32

The Aufbau principle provides a guideline for the order in which orbitals fill. It is a general guideline, but several exceptions occur for specific elements.

Orbitals fill based on their relative energy, with orbitals increasing in energy as *n* increases and as λ increases within a level *n*. Orbitals within a sublevel are equal in energy (degenerate).

The usual order of energies is as shown below. Remember, exceptions to this order do occur.

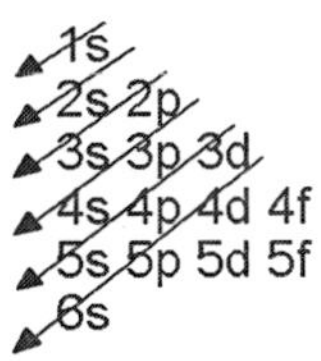

Pauli Exclusion Principle: No two electrons in an atom may have identical sets of four quantum numbers.

Hund's Rule: Electrons must occupy all orbitals of a given sublevel before electron pairing begins. These unpaired electrons have parallel spins.

ORBITAL NOTATIONS

Note electron configurations and orbital notations mostly follow the Aufbau principle.

- Note exceptions to expected order of filling in Cr and Cu - half-filled and filled sets of equivalent orbitals have a special stability.
- Noble gases are very unreactive and have ns^2np^6 configurations (except He) - often oversimplified as "a full outer shell"

Identity	1s	2s	2p	Electron configuration	Simplified configuration
H	↑	↓ ↓	↓ ↓	$1s^1$	$1s^1$
He	↑↓	↓ ↓	↓ ↓	$1s^2$	$1s^2$
Li	↑↓	↓ ↑	↓ ↓	$1s^22s^1$	$[He]2s^1$
Be	↑↓	↓ ↑↓	↓ ↓	$1s^22s^2$	$[He]2s^2$
B	↑↓	↓ ↑↓	↓ ↑↓↓↓	$1s^22s^22p^1$	$[He]2s^22p^1$
C	↑↓	↓ ↑↓	↓ ↑ ↑↓↓	$1s^22s^22p^2$	$[He]2s^22p^2$
N	↑↓	↓ ↑↓	↓ ↑ ↑ ↑	$1s^22s^22p^3$	$[He]2s^22p^3$
O	↑↓	↓ ↑↓	↓ ↑↓ ↑ ↑	$1s^22s^22p^4$	$[He]2s^22p^4$
F	↑↓	↓ ↑↓	↓ ↑↓↑↓↑	$1s^22s^22p^5$	$[He]2s^22p^5$
Ne	↑↓	↓ ↑↓	↓ ↑↓↑↓ ↑↓	$1s^22s^22p^6$	$[He]2s^22p^6$

There is a direct correlation between electron arrangement and arrangement of elements in the periodic table.

- The main group elements contain elements in which the s and p orbitals are being filled.
- For the main group elements, the element's position in the periodic table can be used to determine the number of valence electrons the element has.
- The transition and inner transition metals contain elements in which the highest s level contains 1 or 2 electrons and the d and f orbitals are being filled.

PARAMAGNETISM AND DIAMAGNETISM

- Substances that contain unpaired electrons are paramagnetic (weakly attracted into a magnetic field).

- Substances that have all electrons paired are diamagnetic (very weakly repelled by a magnetic field).
- Which of the following would be paramagnetic - Mg, S, Br, Au, Zn?

TRENDS IN THE PERIODIC TABLE

Chemical Properties

Elements with the same number of valence electrons have similar chemical properties. (Example: Na and K)

Atomic Radius

Increases as we move down in a group because valence electrons are located at higher energy levels, further from the nucleus Decreases as we move from left to right in a period.

- Electrons in inner levels screen/shield electrons in outer levels so that they "feel" less of the nucleus' charge - the effective nuclear charge on these electrons is less than the actual nuclear charge.
- Note: effective nuclear charge = actual nuclear charge - number of electrons in inner levels

What effective nuclear charge do the valence electrons in an aluminum atom "feel"?

In a phosphorous atom?

In a sulphur atom?

- As we move from left to right in a group, the effective nuclear charge increases.
- Increasing effective nuclear charge pulls electrons in more tightly (closer to the nucleus) resulting in a smaller atomic radius.

Onization Energy

First ionization energy - amount of energy required to remove the most loosely bound electron from an isolated gaseous atom

- Ex. $Ca_{(g)}$ + 590 kJ → $Ca^+_{(g)}$ + e^-
  Second ionization energy - amount of energy required to remove the second electron
- Ex. $Ca^+_{(g)}$ + 1154 kJ → $Ca^{2+}_{(g)}$ + e^-
  Second ionization energy is always larger than first ionization energy.
  Higher ionization energies (IE) relate to more tightly bound electrons.
- Elements with low IE lose electrons easily to form cations; elements with high IE gain electrons easily to form anions.
- IE decreases as we move down in a group - most loosely bound electron is further from nucleus.
- IE increases as we move left to right in a period - effective nuclear charge is increasing and atomic radius is decreasing.

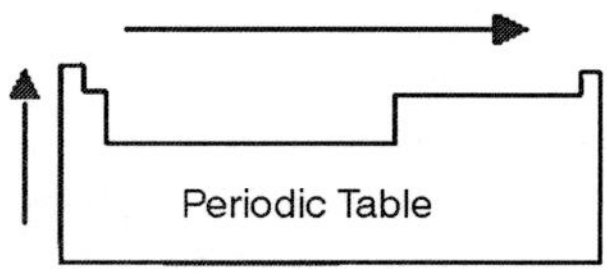

Increasing IE

- Exception 1 - 1st p electron is easier to remove than 2nd s electron because having a filled s sub-level leads to extra stability (compare Be to B).
- Exception 2 - 4th p electron is easier to remove than 3rd p electron because having a half-filled p sub-level leads to extra stability (compare N to O)

Electron Affinity

The amount of energy absorbed when an electron is added to an isolated gaseous atom.

- $Cl_{(g)}$ + e^- → $Cl^-_{(g)}$ + 349 kJ

EA= -349 kJ/mol

Atoms which release energy when an electron is added have negative values for EA - indicating that they easily form anions. General trend in periodic table: EA becomes more negative as we go up in a group and to the right in a period.

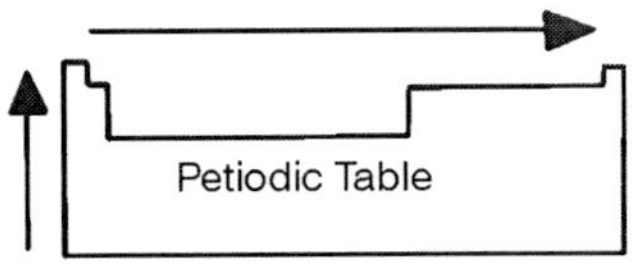

More negative EA

Exceptions occur as noted for ionization energy.

Ionic radius

Ions formed from the loss of electrons are smaller than the neutral atom; ions formed from the gain of electrons are larger than the neutral atom.

- Ex. Li^+ is smaller than Li because it has emptied the second level.
- F- is larger than F because it has an additional electron in its electron cloud, causing the cloud to expand as electrons repel each other.

  Isoelectronic species have the same number of electrons.
- Ex. Al^3+,Mg^2+, Na^+, Ne, F^-, O^2-, N^{3-} all have 10 electrons (2 in 1st level and 8 in 2nd level)

  Sizes within an isoelectronic species vary based on nuclear charge - as nuclear charge increases, the radius decreases.

Electronegativity

A measure of the tendency of an atom to attract electrons to itself when it is chemically combined with another atom

- Elements with high electronegativity values tend to gain electrons (form anions); elements with low electronegativity values tend to lose electrons (form cations)
- When two atoms with similar electronegativity values combine, they are more likely to share electrons.
- Electronegativity values generally increase from left to right and from bottom to top on the periodic table.

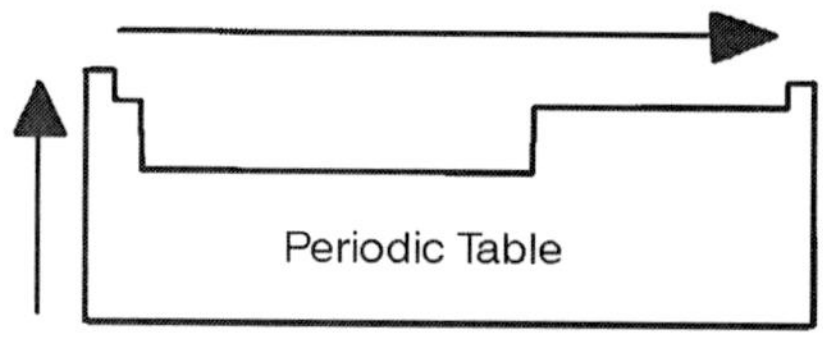

STIRLING'S APPROXIMATION

The ratio of (ln n!) to (n ln n – n) approaches unity as n increases.

In mathematics, Stirling's approximation (or Stirling's formula) is an approximation for large factorials. It is named in honour of James Stirling.

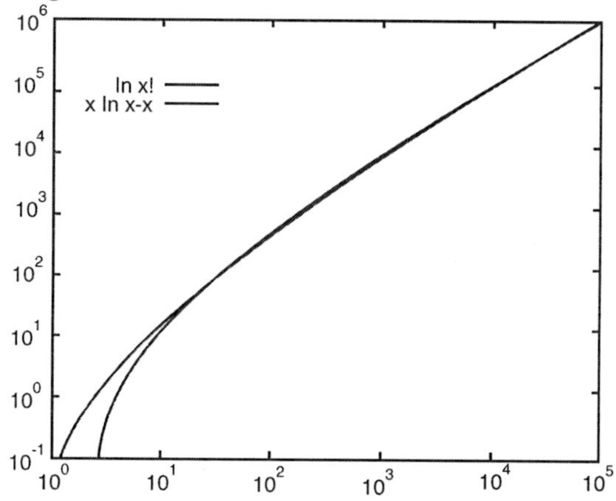

The formula is written as

$$n! \approx \sqrt{2\pi n}\frac{n^n}{e^n}.$$

Roughly, this means that these quantities approximate each other for all sufficiently large integers n. More precisely, Stirling's formula says that

$$\lim_{n\to\infty}\frac{n!}{\sqrt{2\pi n}\frac{n^n}{e^n}}=1.$$

or

$$\lim_{n\to\infty} \frac{e^n n!}{n^n \sqrt{n}} = \sqrt{2\pi}.$$

DERIVATION

The formula, together with precise estimates of its error, can be derived as follows. Instead of approximating n!, one considers the natural logarithm:

$$\text{In}(n!) = \text{In}1 + \text{In}2 + \cdots + \text{Inn}.$$

Then, we can apply the Euler-Maclaurin formula by putting $f(x) = \ln(x)$ to find an approximation of the value of ln(n!).

$$In(n-1)! = nInn - n + 1 + \frac{Inn}{2} + \sum_{k=2}^{m} \frac{B_k(-1)^k}{k(k-1)}\left(\frac{1}{n^{k-1}} - 1\right) + R$$

where B_k is Bernoulli number and R is the remainder of the Euler-Maclaurin formula.

We can then take limits on both sides,

$$\lim_{n\to\infty}\left(Inn! - nInn + n - \frac{Inn}{2}\right) = 1 + \sum_{k=2}^{m} \frac{B_k(-1)^k}{k(k-1)} + \lim_{n\to\infty} R.$$

Let the above limit be y and compound the above two formula, we get the approximation formula in its logarithmic form:

$$In\ n! = \left(n + \frac{1}{2}\right)Inn - n + y + \sum_{k=2}^{m} \frac{B_k(-1)^k}{k(k-1)^{k-1}} + 0\left(\frac{1}{n^m}\right)$$

where $O(f(n))$ is Big-O notation.

Just take the exponential on both sides, and choose any positive integer *m*, say 1. We get the formula with an unknown term ey:

$$n! = e^y \sqrt{n}\left(\frac{n}{e}\right)^n \left(1 + 0\left(\frac{1}{n}\right)\right).$$

The unknown term e^y can be found by taking the limit on both sides as *n* tends to infinity and using Wallis' product. One can approximate the value of e^y by $\sqrt{2\pi}$. Therefore, we get Stirling's formula:

$$n! = \sqrt{2\pi n}\left(\frac{n}{e}\right)^n \left(1 + 0\left(\frac{1}{n}\right)\right).$$

The formula may also be obtained by repeated integration by parts, and the leading term can be found through the method of steepest descent. The general formula (without the n1/2 term) may be quickly obtained by approximating the sum

$$In\ N! = \sum_1^N Inn$$

with an integral:

$$\sum_{n=1}^{N} Inn \approx \int_1^N In\ n\ dn = N\ In N - N + 1.$$

SPEED OF CONVERGENCE AND ERROR ESTIMATES

More precisely,

$$n! = \sqrt{2\pi n}\left(\frac{n}{e}\right)^n e^{\lambda n}$$

with

$$\frac{1}{12n+1} < \lambda_n < \frac{1}{12n}.$$

Stirling's formula is in fact the first approximation to the following series (now called the Stirling series):

$$n! = \sqrt{2\pi n}\left(\frac{n}{e}\right)^n$$

$$\left(1 + \frac{1}{12n} + \frac{1}{288n^2} - \frac{139}{51840n^3} - \frac{571}{2488320n^4} + \ldots\right).$$

As $n \to \infty$, the error in the truncated series is asymptotically equal to the first omitted term. This is an example of an asymptotic expansion.

The asymptotic expansion of the logarithm is also called Stirling's series:

$$Inn! = nInn - n + \frac{1}{2} In(2\pi n)$$

$$+\frac{1}{12n} - \frac{1}{360n^3} + \frac{1}{1260n^5} - \frac{1}{1680n^7} + \ldots$$

In this case, it is known that the error in truncating the series is always of the same sign and at most the same magnitude as the first omitted term.

STIRLING'S FORMULA FOR THE GAMMA FUNCTION

Stirling's formula may also be applied to the Gamma function

$\Gamma(z+1) = \coprod(z) = z!$

defined for all complex numbers other than non-positive integers. If $\Re(z) > 0$ then

$$In\Gamma(z) = \left(x - \frac{1}{2}\right) In\, z - z + \frac{In 2\pi}{2} + 2\int_0^\infty \frac{\arctan \frac{t}{z}}{\exp(x\pi t) - 1} dt.$$

Repeated integration by parts gives the asymptotic expansion

$$In\Gamma(z) = \left(x - \frac{1}{2}\right) In\, z - z + \frac{In 2\pi}{2} + \sum_{n=1}^{\infty} \frac{B_{2n}}{2n(2n-1)z^{2n-1}}$$

where B_n is the nth Bernoulli number. The formula is valid for z large enough in absolute value when $|\arg z| < \pi - \varepsilon$, where ε is positive, with an error term of O(z-m$^{-1/2}$) when the first m terms are used. The corresponding approximation may now be written:

$$\Gamma(z) = \sqrt{\frac{2\pi}{z}} \left(\frac{z}{e}\right)^z \left(1 + 0\left(\frac{1}{z}\right)\right).$$

A CONVERGENT VERSION OF STIRLING'S FORMULA

Thomas Bayes showed, that Stirling's formula did not give a convergent series.

Obtaining a convergent version of Stirling's formula entails evaluating

$$\int_0^\infty \frac{2\arctan\frac{t}{z}}{\exp(2\pi t)-1}dt = In\ \Gamma(z)-\left(z-\frac{1}{2}\right)In\, z+z-\frac{1}{2}In(2\pi).$$

One way to do this is by means of a convergent series of inverted rising exponentials. If, $z^{\bar{n}} = z(z+1)\cdots(z+n-1)$, then

$$\int_0^\infty \frac{2\arctan\frac{t}{z}}{\exp(2\pi t)-1}dt = \sum_{n=1}^{\infty}\frac{c_n}{(z+1)^{\bar{n}}}$$

where

$$c_n = \frac{1}{n}\int_n^1 x^{\bar{n}}\left(x-\frac{1}{2}\right)dx$$

From this we obtain a version of Stirling's series

$$In\ \Gamma(z) = \left(z-\frac{1}{2}\right)In\, z - z + \frac{In 2\pi}{2} + \frac{1}{12(z+1)}$$

$$+\frac{1}{12(z+1)(z+2)} + \frac{59}{360(z+1)(z+2)(z+3)}$$

$$+\frac{29}{90(z+1)(z+2)(z+3)(z+4)}+\ldots$$

which converges when $\Re(Z) > 0$..

A VERSION SUITABLE FOR CALCULATORS

The approximation

$$\Gamma(z) \approx \sqrt{\frac{2\pi}{z}}\left(\frac{z}{e}\sqrt{z\sinh\frac{1}{z}+\frac{1}{810z^6}}\right)^z,$$

or equivalently,

$$2\, In\Gamma(z) \approx In(2\pi) - In\, z$$

$$+z\left(2 In z + In\left(z\sinh\frac{1}{z}+\frac{1}{810z^6}\right)-2\right),$$

can be obtained by rearranging Stirling's extended formula and observing a coincidence between the resultant power series and the Taylor series expansion of the hyperbolic sine function.

This approximation is good to more than 8 decimal digits for z with a real part greater than 8. Robert H. Windschitl suggested it in 2002 for computing the Gamma function with fair accuracy on calculators with limited programme or register memory.

GergöNemes proposed in 2007 an approximation which gives the same number of exact digits as the Windschitl approximation but is much simpler:

$$\Gamma(z) \approx \sqrt{\frac{2\pi}{z}}\left(\frac{1}{e}\left(z+\frac{1}{12z-\frac{1}{10z}}\right)\right)^{z},$$

or equivalently,

$$In\Gamma(z) \approx \frac{1}{2}(In(2\pi) - Inz) + z\left(In\left(z+\frac{1}{12z-\frac{1}{10z}}\right)-1\right).$$

The formula was first discovered by Abraham de Moivre in the form

$$n!`\sim[\text{constant}].n^{n+1/2}e^{-n}.$$

Stirling's contribution consisted of showing that the constant is $\sqrt{2\pi}$. The more precise versions are due to Jacques Binet.

The "first-order" version of Stirling's approximation, $n! \approx n^n$, was used by Max Planck in his 1901 article on the black body radiation formula. It linked Planck's concept of energy elements to the black body radiation formula for very large numbers of energy elements and oscillators. The approximation was often used in quantum theory, for example by Debye and de Broglie. Einstein and Bose took a different approach. For very large n, the graph of the probability expression Planck obtained using the "first order" Stirling's formula, plotted in a logarithmic coordinate system, is almost parallel to the line obtained direct from the idea of separated light quanta.

However, the system entropy calculated by using the "first order" approximation is different and the ratio gets strongly nonlinear for small n. One can only speculate that a similar total effect on entropy could be obtained by introducing the uncertainty principle and the photon spin as well as other quantities which were unknown at the time when the old quantum theory was created. Unfortunately, the experimental verification of the link between the "first order" Stirling's approximation and modern physical theories is still missing.

SUBSYSTEM AND TEMPERATURE OF ISOLATED PARAMAGNET

BETA AND THE TEMPERATURE

A negative temperature coefficient (NTC) thermistor is a two terminal solid state electronic component that exhibits a large, predictable change in resistance corresponding to changes in absolute body temperature. This change in body temperature of the thermistor can be brought about either externally via a change in ambient temperature or internally by heat resulting from current passing through the device or by a combination of these effects.

NTC thermistors are manufactured using metallic oxides of manganese, nickel, cobalt, copper, iron and other metals. They are fabricated using a mixture of two or more metallic oxides and a binder material and are then pressed into the desired configuration.

The resulting material is then sintered at elevated temperatures. By varying the types of oxides, the sintering time and temperature as well as the atmosphere, a wide variety of curves and resistance values can be manufactured.

Thermistor Terminology

Thermistors exhibit a large negative change in resistance with respect to temperature, on the order of -3%/C to -6%/°C at 25°C. This relationship between resistance and temperature follows an approximately exponential-type curve. A few parameters will help to describe the curve and how it changes over temperature.

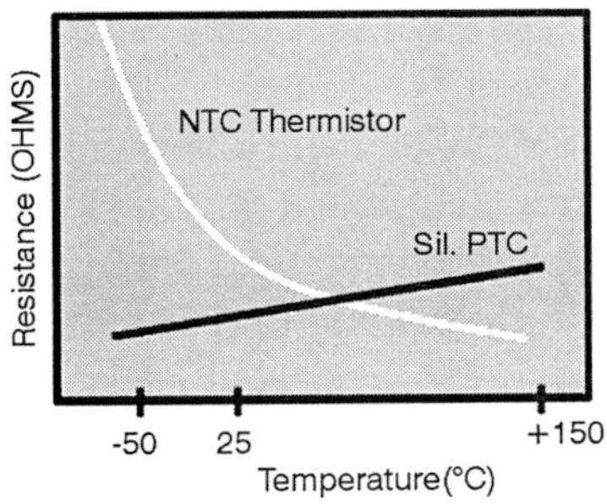

Fig. Resistance vs Temperature Graph

Resistance at 25’IC (R_{25})

The most common temperature used to measure the thermistor resistance and the one temperature that is most often used to reference the resistance value of the thermistor is 25’IC. For NTC thermistors, this value can vary from less than 100 to greater than 1Meg.

The value at 25’IC is normally measured in a temperature controlled bath where very low power is used to measure the resistance value. When a resistance value for a thermistor is mentioned, it is the value at 25’IC that is usually being used.

Temperature Coefficient of Resistance (α)

One way to describe the curve of an NTC thermistor is to measure the slope of the resistance versus temperature (R/T) curve at one temperature. By definition, the coefficient of resistance is given by:

$$\alpha = \frac{1}{R} * \frac{dR}{dT}$$

where:

- T = Temperature in °C or K
- R = Resistance at Temp T

The temperature coefficient is expressed in ohms/ohms/IC or more commonly %/IC.

The steepest portion of the NTC curve is at colder temperatures. Depending upon the type of NTC material, the temperature coefficient at -40°C can be as high as -8%/°C. The flattest portion of the curve occurs at higher temperatures where, at temperatures of 300°C, a can be less than 1%/°C.

The temperature coefficient is one method that can be used to compare the relative steepness of NTC curves. It is important that the temperature coefficient be compared at the same temperature because, as was noted previously, a varies widely over the operating temperature range.

Resistance Ratio (Slope)

The resistance ratio, or slope, for thermistors is defined as the ratio of resistance at one temperature to the resistance at a second higher temperature. The resistance ratio is one method of describing the NTC curve. It is sometime used to compare the relative steepness of two curves. There is no industry standard for the two temperatures that are used to calculate the ratio, although some common temperature ranges are:

$$\frac{R@0°C}{R@50°C} \; \frac{R@25°C}{R@50°C} \; \frac{R@25°C}{R@85°C}$$

The value obtained by taking the resistance ratio at different temperatures will vary greatly depending upon the temperatures used.

Therefore, resistance ratios cannot be used to compare thermistor curves unless the same temperature ranges are used. For ATP Curve "Z", the following ratios are obtained:

$$\frac{R@0°C}{R@50°C} \frac{3.265}{0.3601} = 9.07$$

$$\frac{R@25°C}{R@50°C} \frac{1.000}{0.3601} = 2.78$$

$$\frac{R@25°C}{R@85°C} \frac{1.000}{0.1071} = 9.34$$

Beta Value (β)

A simple approximation for the relationship between the resistance and temperature for a NTC thermistor is to use an exponential approximation between the two. This approximation is based on simple curve fitting to experimental data and uses two points on a curve to determine the value of. The equation relating resistance to temperature using is:

$R = Ae^{(\beta/T)}$

Where:

R = thermistor resistance at temp T

A = constant of equation

β = Beta, the material constant

T = Thermistor temperature (K)

To calculate Beta for any given temperature range, the following formula applies:

$$\beta = \left(\frac{T_1 k * T_2}{T_2 - T_1}\right) In \frac{R_1}{R_2}$$

β can be used to compare the relative steepness of NTC thermistor curves. However, as with resistance ratios, the value of β will vary depending upon the temperatures used to calculate the value, although not to the extent that resistance ratio does. For example, to calculate β for the temperature range of 0°C to 50°C for ATP curve "Z":

T1 = 0°C + 273.15°C = 273.15K

T2 = 50°C + 273.15°C = 323.15K

R_1 = 3.265

R_2= 0.3601

This value of β would be referenced as β0°C/50°C. Using other temperatures to calculate b for curve "Z" would yield the following results:

$\beta_{25°C/50°C}$ = 3936K

$\beta_{25°C/85°C}$ = 3976K

As we can see, it is important to know what temperatures were used to calculate the value of β before it is used to compare thermistor curves. b can be used to calculate the resistance of the curve at other temperatures within the range that b was calculated once the constant A is determined. However, the accuracy of this equation is only approximately ±0.5°C over a 50°C span.

Steinhart-Hart Thermistor Equation

The Steinhart-Hart equation is an emperically derived polynomial formula which best represents the resistance versus temperature relationship of NTC thermistors. The Steinhart-Hart equation is the best method used to describe the RvT relationship and is accurate over a much wider range of temperature than is. To solve for temperature when resistance is known, yields the following form of the equation:

$$1/T = a + b(LnR) + c(LnR)^3$$

Where:

T = temperature in Kelvins (K = °C + 273.15)

a, b and c are equation constants

R = resistance in Ω at temp T

To solve for resistance when the temperature is known, the form of the equation is:

$$R = e^{\left[\left|-\frac{X}{2}+\frac{X^3}{4}+\frac{\psi}{27}\right|^{'1b} + \left|-\frac{X}{2}-\frac{X^3}{4}+\frac{\psi'}{27}\right|^{1b}\right]}$$

Where:

$$X = \frac{a-1/T}{C} \qquad \psi = \frac{b}{C}$$

The a, b and c constants can be calculated for either a thermistor material or for individual values of thermistors within a material type. To solve for the constants, three sets of data must be used. Normally, for a temperature range, values at the low end, middle and high end are used to calculate the constants. This will ensure the best fit for the equation over the range. Using the Steinhart-Hart equation allows for an accuracy as good as ±0.001°C over a 100°C temperature span.

THERMISTOR TOLERANCE AND TEMPERATURE ACCURACY

There are two factors to consider when discussing thermistors and their ability to measure temperature. The first is resistance tolerance and this is defined as the amount of resistance that any part will vary from its nominal value. The tolerance on the resistance at any temperature is the sum of:

- The closest tolerance at any specified temperature
- The additional tolerance due to deviation from the nominal curve for the material

In any application where the thermistor is to be used to measure temperature it is more appropriate to discuss the temperature accuracy for the device. The accuracy can be calculated if the resistance tolerance and α are known.

There are two generally accepted methods of describing the tolerance or accuracy of a thermistor. The first is point matched. This describes a thermistor that has its tightest resistance tolerance at one temperature, the reference temperature, which is normally 25°C. At temperatures below and above the reference temperature the resistance tolerance will become larger due to the uncertainty in the material curve. The other type of thermistor tolerance is known as curve matched or interchangeable. These thermistors are normally defined to have a certain accuracy over a range, typically ±0.2°C from 0°C to 70°C.

A simple equation is used to describe the relationship between resistance tolerance and temperature accuracy. When one is known the other can be calculated.

$$Accy = \frac{\text{Res Tol}}{\alpha} \quad or \quad \text{Res Tol=Accy}\bullet\alpha$$

For example, for ATP part number A1004Z-2, the resistance tolerance is ±2% @ 25°C. Looking at the data for curve "Z" shows that the a at 25°C is 4.4 %/°C. Therefore, the accuracy at 25°C can be calculated to be (±2% / 4.4%/°C) = ±0.45°C.

Similarly, for ATP part number A1004Z-C3, the temperature accuracy is expressed as ±0.2°C from 0°C to 70°C. To calculate the resistance tolerance at 25°C divide the temperature accuracy at the temperature by the a at that temperature. For 25°C, the resistance tolerance would be (±0.2°C * 4.4%/°C) = ±0.88%.

In the data section for NTC thermistors, ATP also provides the curve deviation for parts that are point matched at 25°C. Using this information and the value of, allows for the temperature accuracy to be calculated at any temperature. For example for curve "Z" at 50°C for ATP part number A1004Z-2, the resistance tolerance at 25°C is ±2%. The deviation due to the curve uncertainty is listed as ±1.2%. Therefore, the total resistance tolerance would be:

(±2%) + (±1.2%) = ± 3.2% @ 50°C

The at 50°C for this material is listed as -3.8%/°C. Therefore, to calculate the temperature accuracy at 50°C for A1004Z-2:

(±3.2%) / (-3.8%/°C) = ±0.84°C

The a at 50°C for this material is listed as -3.8%/°C.

NTC Thermistor Self-heated Parameters

Self-heating occurs in a thermistor when current passing through the device is such that the internal heat generated is sufficient to raise the thermistor body temperature above that of its environment.

For temperature sensing applications, it is not desirable to self-heat the thermistor to any extent. Other NTC thermistor applications utilize the self heated characteristics inherent to the parts. The ability of a thermistor to dissipate power is a function of the size of the part, its geometry, lead material and size, method of mounting and any other factor that would contribute to the ability of the part to dissipate heat.

Dissipation Factor (δ)

The dissipation factor, δ, defines the relationship between the applied wattage and the thermistor self heating in terms of temperature rise. This relationship is defined as follows:

$$\delta = \frac{P}{\Delta T}$$

where:

P = power dissipated in watts

ΔT = the rise in temperature (°C)

The dissipation factor (δ) is expressed in units of mW/°C. A particular value of δ will correspond to the amount of power necessary to raise the body temperature of the thermistor by 1°C. Because the dissipation factor, δ, is dependent upon a number of factors, the values listed in the data sheets are for reference only.

Time Constant (τ)

The thermal time constant for a thermistor is defined as the time required for a thermistor to change 63.2% of the difference between the initial temperature of the thermistor and that of its surroundings when no power is being dissipated by the thermistor. The value of defines a response time for the thermistor when it has been subjected to a step change in temperature.

A thermistor that has been in an ambient temperature of 25°C for a period of time long enough for it to reach equilibrium, is then moved to an environment where the temperature is 75°C. The thermistor will not immediately indicate a resistance corresponding to the new temperature but rather will exponentially approach the new resistance value. For measurement will correspond to 63.2% of the temperature span, i.e.

$$T_\tau = 0.632\ (70\text{-}25) = 31.6 + 25 = 56.6°C$$

Therefore the temperature that the part must reach is 56.6°C. The resistance of the part at that temperature can be calculated using the Steinhart-Hart equation or an approximation can be used. For example for ATP part number A1004Z-C3, using the equation, the value at 56.6°C should be 2814_Ω. Therefore, to find the value for, τ, we would monitor the resistance value of the part using a multimeter or similar instrument.

The part should start at 25°C where the resistance should be $10{,}000_\Omega$. The time that the part takes to reach 2814_Ω once the part is moved to the new temperature of 75°C will correspond the value of τ and will have the units of seconds. The factors that affect τ are similar to those that affect δ and include the mass of the thermistor, mounting, environment and other factors.

NEGATIVE TEMPERATURE

In physics, certain systems can achieve negative temperatures; that is, their thermodynamic temperature can be of a negative quantity. Negative temperatures can be expressed as negative numbers on the Kelvin scale. Temperatures that are expressed as negative numbers on the familiar Celsius or Fahrenheit scales are simply colder than the zero points of those scales. By contrast, a system with a truly negative temperature is not colder than absolute zero; in fact, temperatures colder than absolute zero are impossible. Rather, a system with a negative temperature is hotter than the same system with an infinite temperature.

Heat and Molecular Energy Distribution

Negative temperatures can only exist in a system where there are a limited number of energy states. As the temperature is increased on such a system, particles move into higher and higher energy states, and as the temperature becomes infinite, the number of particles in the lower energy states and in the higher energy states becomes equal. (This is a consequence of the definition of temperature in statistical mechanics for systems with limited states.)

By injecting energy into these systems in the right fashion, it is possible to create a system in which there are more particles in the higher energy states than in the lower ones. This situation can be characterised as having a negative temperature. A substance with a negative temperature is not colder than absolute zero, but rather it is hotter than infinite temperature. As Kittel and Kroemer put it, "The temperature scale from cold to hot runs +0 K,..., +300 K,..., +∞ K, —∞ K,..., –300 K,..., –0 K."

Generally, temperature as it is felt is defined by the kinetic energy of atoms (heat). Since there is no upper bound on momentum of an atom there is no upper bound to the number of energy states available if enough energy is added, and no way to get to a negative temperature. However, temperature is more generally defined by statistical mechanics than just kinetic energy.

TEMPERATURE AND DISORDER

The distribution of energy among the various translational, vibrational, rotational, electronic, and nuclear modes of a system determines the macroscopic temperature. In a "normal" system, thermal energy is constantly being exchanged between the various modes.

However, for some cases it is possible to isolate one or more of the modes. In practice the isolated modes still exchange energy with the other modes, but the time scale of this exchange is much slower than for the exchanges within the isolated mode. One example is the case of nuclear spins in a strong external magnetic field. In this case, energy flows fairly rapidly among the spin states of interacting atoms, but energy transfer between the nuclear spins and other modes is relatively slow. Since the energy flow is predominantly within the spin system, it makes sense to think of a spin temperature that is distinct from the temperature due to other modes.

A definition of temperature can be based on the relationship:

$$T = \frac{dq_{rev}}{ds}$$

The relationship suggests that a positive temperature corresponds to the condition where entropy, *S*, increases as thermal energy, q_{rev}, is added to the system. This is the "normal" condition in the macroscopic world and is always the case for the translational, vibrational, rotational, and non-spin related electronic and nuclear modes. The reason for this is that there are an infinite number of these types of modes and adding more heat to the system increases the number of modes that are energetically accessible, and thus the entropy.

Nuclear Spins

In the case of electronic and nuclear spin systems there are only a finite number of modes available, often just two, corresponding to spin up and spin down. In the absence of a magnetic field, these spin states are degenerate, meaning that they correspond to the same energy. When an external magnetic field is applied, the energy levels are split, since those spin states that are aligned with the magnetic field will have a different energy than those that are anti-parallel to it.

In the absence of a magnetic field, one would expect such a two-spin system to have roughly half the atoms in the spin-up state and half in the spin-down state, since this maximizes entropy. Upon application of a magnetic field, some of the atoms will tend to align so as to minimize the energy of the system, thus slightly more atoms should be in the lower-energy state (for the purposes of this example we'll assume the spin-down state is the lower-energy state). It is possible to add energy to the spin system using radio frequency (RF) techniques. This causes atoms to flip from spin-down to spin-up.

Since we started with over half the atoms in the spin-down state, initially this drives the system towards a 50/50 mixture, so the entropy is increasing, corresponding to a positive temperature. However, at some point more than half of the spins are in the spin-up position. In this case, adding additional energy reduces the entropy since it moves the system further from a 50/50 mixture. This reduction in entropy with the addition of energy corresponds to a negative temperature.

Semiconductor Lasers

This phenomenon can also be observed in many lasing systems, wherein a large fraction of the system's atoms (for chemical and gas lasers) or electrons (in semiconductor lasers) are in excited states. This is referred to as a population inversion.

The Hamiltonian for a single mode of a luminescent radiation field at frequency v is

$$H = (hv - \mu)a.$$

The density operator in the grand canonical ensemble is

$$\rho = \frac{\exp(-\beta H)}{Tr\,\exp(-\beta H)}.$$

For the system to have a ground state, the trace to converge, and the density operator to be generally meaningful, βH must be positive semidefinite. So if $hv < \mu$ and H is negative semidefinite, then β must itself be negative, implying a negative temperature.

Negative Temperature Coefficient

A negative temperature coefficient (NTC) occurs when the thermal conductivity of a material rises with increasing temperature, typically in a defined temperature range. For most materials, the thermal conductivity will decrease with increasing temperature.

Materials with a negative temperature coefficient have been used in floor heating since 1971. The negative temperature coefficient avoids excessive local heating beneath carpets, bean bag chairs, mattresses etc., which can damage wooden floors, and may infrequently cause fires.

Most ceramics exhibit NTC behaviour, which is governed by an Arrhenius equation over a wide range of temperatures: R = A exp (B/T) where R = resistance A, B = constants T = absolute temperature (K) The constant B is related to the energies required to form and move the charge carriers responsible for electrical conduction - hence, as the value of B decreases, the material becomes insulating. Practical and commercial NTC resistors aim to combine modest resistance with a value of B that provides good sensitivity to temperature.

Such is the importance of the B constant value, that it is possible to characterize NTC thermistors using the B parameter equation: R = r∞eB/T where r∞ = R0 e-B/T0 R0 = resistance at temperature T0 Therefore, many materials that produce acceptable values of B include materials that have been alloyed or possess variable cation valence states and thus contain a high natural defect centre concentration. The value of B strongly depends on the energy required to dissociate the charge carriers that are used for the electrical conduction from these defect centres.

NTC thermistors are generally manufactured from pressed die chip of semi-conducting material - often a sintered ceramic oxide - and are based on a conduction model. A semi conductor is intermediate between an insulator and a conductor, and behaves as an insulator as low temperatures and becomes more conducting as temperature increases.

One of the main reasons that semi-conductors are so useful for thermistors is that their electrical properties can be controlled and enhanced by 'doping' or alloying with impurities - either elements or compounds.

A common commercially available NTC thermistor material is Mn_3O_4 which can be doped with varying amounts of NiO. Since the N_i^{2+} cation has less positive charges than the Mn^{3+} cation so to maintain charge neutrality a M_n^{3+} is converted to a Mn^{4+} cation for each of the substituted Ni^{2+} cations.

The electrical conduction is therefore increase by an electron hopping mechanism between the M_n^{3+} and Mn^{4+} ions on equivalent sites in the crystal lattice. It is also possible to dope the Mn_3O_4 with other combinations of impurities such as CoO and CuO. Due to these properties of semi-conductors, as the temperature increases, the energy of the electrons increases due to the increased thermal energy available and therefore enables them to be promoted to the conduction band. This results in a greater proportion of electrons being able to move around and carry charge - the more electrons that are mobile and can carry charge, the more current a material can conduct and therefore the resistivity has decreased - hence the NTC thermistors exhibit a decrease in resistivity with increasing temperature. This phenomenon is also described by the change in the temperature coefficient of resistance: $\alpha = (1/R)\,dR/dT$ $\alpha = -B/T_3$. It is therefore clear that the temperature coefficient of resistance is inversely proportion to the temperature, and decreases when the temperature increases.

Negative Temperature Coefficient of Reactivity

In a nuclear reactor, temperature changes can introduce reactivity changes. This property is called the "temperature coefficient of reactivity." In water-cooled nuclear reactors, the predominant reactivity changes are brought about by changes in the temperature of the coolant water. In this case the temperature coefficient is negative, which means that an increase in coolant temperature causes a decrease in reactivity, and vice-versa. A reactor with a negative temperature coefficient of reactivity is therefore inherently self-controlling and safe.

NEGATIVE TEMPERATURE MEAN

Under certain conditions, a closed system can be described by a negative temperature, and, surprisingly, be hotter than the same system at any positive temperature.

Temperature

To get things started, we need a clear definition of

"temperature." Actually various kinds of "temperature" appear in the literature of physics (e.g., kinetic temperature, colour temperature). The relevant one here is the one from thermodynamics, in some sense the most fundamental.

Our intuitive notion is that two systems in thermal contact should exchange no heat, on average, if and only if they are at the same temperature. Let's call the two systems S1 and S2. The combined system, treating S1 and S2 together, can be S3. The important question, consideration of which will lead us to a useful quantitative definition of temperature, is "How will the energy of S3 be distributed between S1 and S2?"

With a total energy E, S has many possible internal states (microstates). The atoms of S3 can share the total energy in many ways. Let's say there are N different states. Each state corresponds to a particular division of the total energy in the two subsystems S1 and S2. Many microstates can correspond to the same division, E1 in S1 and E2 in S2.

A simple counting argument tells us that only one particular division of the energy, will occur with any significant probability. It's the one with the overwhelmingly largest number of microstates for the total system S3. That number, N(E1,E2) is just the product of the number of states allowed in each subsystem, N(E1,E2) = N1(E1)*N2(E2), and, since E1 + E2 = E, N(E1,E2) reaches a maximum when N1*N2 is stationary with respect to variations of E1 and E2 subject to the total energy constraint.

For convenience, physicists prefer to frame the question in terms of the logarithm of the number of microstates N, and call this the entropy, S.

We can easily see from the above analysis that two systems are in equilibrium with one another when $(dS/dE)_1 = (dS/dE)_2$, i.e., the rate of change of entropy, S, per unit change in energy, E, must be the same for both systems. Otherwise, energy will tend to flow from one subsystem to another as S3 bounces randomly from one microstate to another, the total energy E3 being constant, as the combined system moves towards a state of maximal total entropy. We define the temperature, T, by $1/T = dS/dE$, so that the equilibrium condition becomes the very simple $T_1 = T_2$.

This statistical mechanical definition of temperature does in fact correspond to our intuitive notion of temperature for most systems. So long as dS/dE is always positive, T is always positive. For common situations, like a collection of free particles, or particles in a harmonic oscillator potential, adding energy always increases the number of available microstates, increasingly faster with increasing total energy.

So temperature increases with increasing energy, from zero, asymptotically approaching positive infinity as the energy increases.

Negative Temperature

Not all systems have the property that the entropy increases monotonically with energy. In some cases, as energy is added to the system, the number of available microstates, or configurations, actually decreases for some range of energies. Imagine an ideal "spin-system", a set of N atoms with spin 1/2 on a one-dimensional wire.

The atoms are not free to move from their positions on the wire. The only degree of freedom allowed to them is spin-flip: the spin of a given atom can point up or down.

The total energy of the system, in a magnetic field of strength B, pointing down, is $(N^+ - N^-)*uB$, where u is the magnetic moment of each atom and N^+ and N^- are the number of atoms with spin up and down respectively. Notice that with this definition, E is zero when half of the spins are up and half are down. It is negative when the majority are down and positive when the majority are up.

The lowest possible energy state, all the spins pointing down, gives the system a total energy of -NuB, and temperature of absolute zero. There is only one configuration of the system at this energy, i.e., all the spins must point down. The entropy is the log of the number of microstates, so in this case is log(1) = 0. If we now add a quantum of energy, size uB, to the system, one spin is allowed to flip up. There are N possibilities, so the entropy is log(N). If we add another quantum of energy, there are a total of N(N – 1)/2 allowable configurations with two spins up. The entropy is increasing quickly, and the temperature is rising as well. However, for this system, the entropy does not go on increasing forever. There is a maximum energy, +NuB, with all spins up. At this maximal energy, there is again only one microstate, and the entropy is again zero. If we remove one quantum of energy from the system, we allow one spin down. At this energy there are N available microstates. The entropy goes on increasing as the energy is lowered. In fact the maximal entropy occurs for total energy zero, i.e., half of the spins up, half down.

So we have created a system where, as we add more and more energy, temperature starts off positive, approaches positive infinity as maximum entropy is approached, with half of all spins up. After that, the temperature becomes negative infinite, coming down in magnitude toward zero, but always negative, as the energy increases toward maximum. When the system has negative temperature, it is hotter than when it is has positive temperature. If we take two copies of the system, one with positive and one with negative temperature, and put them in thermal contact, heat will flow from the negative-temperature system into the positive-temperature system.

Atoms always have other degrees of freedom in addition to spin, usually making the total energy of the system unbounded upward due to the translational degrees of freedom that the atom has. Thus, only certain degrees of freedom of a particle can have negative temperature.

It makes sense to define the "spin-temperature" of a collection of atoms, so long as one condition is met: the coupling between the atomic spins and the other degrees of freedom is sufficiently weak, and the coupling between atomic spins sufficiently strong, that the timescale for energy to flow from the spins into other degrees of freedom is very large compared to the timescale for thermalization of the spins among themselves. Then it makes sense to talk about the temperature of the spins separately from the temperature of the atoms as a whole. This condition can easily be met for the case of nuclear spins in a strong external magnetic field.

Nuclear and electron spin systems can be promoted to negative temperatures by suitable radio frequency techniques. Various experiments in the calorimetry of negative temperatures, as well as applications of negative temperature systems as RF amplifiers, etc., and the references therein.

IS A NEGATIVE KELVIN TEMPERATURE POSSIBLE?

Thermodynamics is older than the atomistic view of matter, hence thermodynamic quantities such as temperature are not defined atomistically. The definition of temperature allows negative absolute temperatures. But do such temperature have any empirical meaning? While this might be questionable, it is certainly interesting to discuss it because it helps understand the concept of temperature. As follows, a simplified statistical concept of temperature is discussed and used to explain the state of matter in lasers as well as the technique to attain very low temperature (adiabatic demagnetization)

The most often given answer to this question is something like "no, because at zero K, molecules are not moving anymore and we cannot have less than zero motion."

This answer, while not completely wrong, ignores that temperature is not defined as motion of molecules. It is, for certain systems, equal to the mean kinetic energy of molecules, but this is an equality, not a definition. Temperature is usually taken as simple, because we all can feel it. But have we ever tought of explaining it to an alien who does not feel it? It would be very difficult. Entropy, for instance, is much simpler, yet seems difficult, probably because we cannot feel it. What is correct in the above-cited answer is that negative Kelvin temperature cannot be achieved by cooling; moreover, not even zero K can be achieved by cooling, as we probably know. Let's have a closer look at that equality between temperature and mean kinetic energy of molecules, and see what we can get out of this.

Statistical Model

In a gas made of tiny hard billard balls temperature is proportional to mean kinetic energy of the particles. But certainly, nobody expects all molecules to have the same speed. The image (Maxwellian distribution) shows the distribution of molecular speeds for three different temperatures.

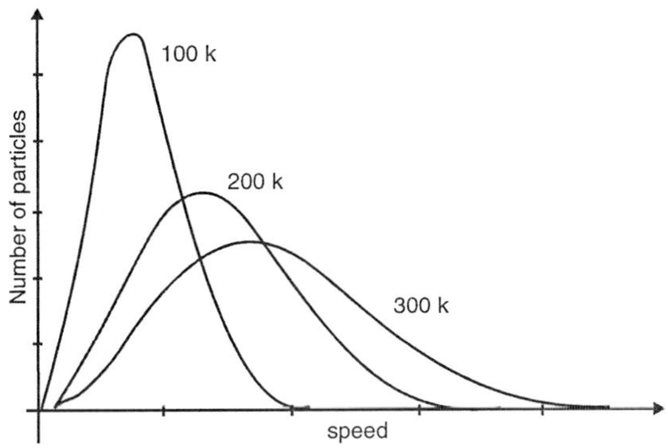

Maxwellian distribution of particle speeds

The Maxwellian distribution of molecular speeds for three different temperatures.

Even at low temperature, there is a small fraction of molecules having high speed. This fraction increases with temperature, while the fracition of molecules with low speed becomes smaller but does not vanish

Obviously, the distribution of molecular speeds depends on temperature. Conversely, temperature is determined by the ratio of the number of molecules at high speed to the number of molecules at low speed. A model can help us understand this relationship.

To keep things simple, a system is assumed that can have but two levels of energy, E(hi) and E(lo): the particles either have energy or they have not.

When the system is heated, energy is transferred to it; the particles must somehow accomodate this energy. In this model, this can only happen if some change from a low to a high energy state.

What is the number of particles that will be on the i-th energy level? The answer is given by the Boltzmann distribution:

N(i) = C*exp (-E(i)/kT),

where

C a constant (at a given temperature)

N(i) the number of particles with energy E(i)

E(i) is the energy portion, according to our simplifying assumption it is either nought or it is something

k the Boltzmann constant

T absolute temperature

For only two levels of energy the ratio of the population of these levels is:

N(hi)/N(lo) = exp (-DeltaE/kT)

where

DeltaE = E(hi)-E(lo)

Solving this equation for T

- DeltaE/k

T = ln [N(hi)/N(lo)]

HEATING AND COOLING

If heat is added, more and more particles change from the low energy level to the high energy level (note that the levels themselves are unchanged); the number of particle with high energy, N(hi), grows and N(lo) decreases so the logarithm becomes less negative and temperature rises. The energy of the whole ensemble rises because now there are more particles on the high energy level. Try calculating the temperature for the system shown in picture (two level model), but one particle changed his place from low to high level.

On the other hand, if energy is withdrawn from a body, E(lo), the low energy level, is populated at the cost of E(hi). The value of the logarithm becomes more and more negative

$$\lim_{\frac{N(hi)}{N(10)} \to \theta} \left[\frac{-\Delta E / k}{In[\frac{N(hi)}{N(lo)}]} \right] = \theta$$

and temperature goes towards zero. Also, from the expression can be concluded that it must be difficult to reach absolute zero, as a diagram of the logarithm versus the ratio shows.

INFINITE TEMPERATURE

System At infitine Temperature

At equally populated levels, temperature, but not the energy, of a system is infinite

If both levels are equally populated, N(hi) = N(lo) and the logarithm vanishes, so that

$$\lim_{N(hi)\to N(\mathrm{lo})}\left[\frac{-\Delta E/k}{In[\frac{\mathrm{N(hi)}}{\mathrm{N(lo)}}]}\right]=\infty$$

The value of the tempera-ture becomes infinite (note that DeltaE does not change). Since number of particles are always finite, only an finite amount of energy is needed to get infinite temperature in this model. Infinite temperature cannot be reached anyway by heating, although not because of the energy needed, but because heat flows from high temperature to low temperature. To heat a body, a hotter body is required.

NEGATIVE ABSOLUTE TEMPERATURE

System at Negative Absolute Temperature

If more particles are at the upper level than at the lower one, absolute temperature of a system is negative

As soon as the high energy level is populated more that the low energy one, we have negative absolute temperature. Can such a state be realised?

Yes, it can, but not by cooling. The formula, as well as the picture, show that a state of matter to which negative absolute tempera-ture can be attributed has more energy than the states at usual temperatures, because more particles are at high energy level than at low energy level.

Thus one has to add energy to get negative absolute temperature. It has been emphasized that such states cannot be reached by adding heat to a body.

Systems at Negative Absolute Temperature

Nevertheless, both systems at infinite temperature and at negative absolute temperature can be prepared. Let us look first at negative temperature. In most lasers atoms are excited electronically to a high energy level. Prior to laser light emission of the whole system, more atoms have to be excited than are not, and a negative absolute temperature can be ascribed to the system. Note that, as expected, atoms are not excited by heating.

Systems at Infinite Temperature

Systems at infinite temperature are used to attain the lowest temperatures possible. The technique is that of adiabatic demagnetization. To get a rough idea of the subject, let us look at magnetism due to electron spin. Each electron can be seen as a tiny magnet, except that it can have only two orientations in an external magnetic field, namely parallel to the field and antiparallel to it. Without a field, these "electron magnets" are randomly oriented. Energy of both states is equal, so it is equally probable to find each orientation. If an external field is applied, the energy of the parallel state is lowered and energy of the antiparallel state goes up.

This phenomenon can be used to cool samples of suitable materials (specific magnetic properties are required). A sample is cooled down to the temperature of liquid helium. Still submerged in liquid helium, a magnetic field is applied to it. At the very first moment, the number of atoms in the parallel equals that in the antiparallel state. But these states are no more energetically equivalent in the presence of an external magnetic field.

Our formula says that if two states differ in energy but are equally populated, then the temperature is infinite. This is an example for the previously stated assertion that one does not need infinite energy to get infinite temperature. In fact, the energy of the sample does not rise, on the contrary! Because it is at infinite temperature, the system will now lose energy as antiparallel states change to parallel. The energy is transferred as heat from the sample to the helium bath. Its temperature changes from infinity to that of the helium bath.

Now the sample is insulated from the helium bath and the external magnetic field is reduced to zero. The energy that was transferred to the helium bath cannot go back to the sample. Hence in this step, while the populations of the antiparallel and the parallel states are gradually equalized, the sample cools down. This is how some of the lowest available temperatures are attained.

SYSTEMS WITH MORE THAN TWO ENERGY LEVELS

Usual systems have more than two excited states and hence more than two energy levels. Can these temperatures be attributed to such systems, too?

Looking at the limits of this two level model may help to elucidate further the concept of temperature. First, each level must be populated enough to avoid random fluctuations of particle number. Keep in mind that the systems are dynamic, particles exchanging energy continually. So the model fails when there are only a few particles on each level.

The above given Boltzmann relation holds true for multi level systems, of course. Now we can clarify what it means when it is said that a system must be in thermal equilibrium:

A system is in thermal equilibrium if particle number on each energy level follows the Boltzmann distribution.

Random fluctuations cause deviations from this distribution, but the larger the fluctuation, the less it is probable, so most of the time the system will be close to that distribution.

For systems with an infinite number of states infinite temperature would also mean infinite energy. Now the relation

$Ekin = {}^{3}/_{2}kT$

At equally populated levels, temperature, but not the energy, of a system is infinite

If both levels are equally populated, N(hi) = N(lo) and the logarithm vanishes, so that

$$\lim_{N(hi)\to N(\text{lo})}\left[\frac{-\Delta E/k}{In[\frac{\text{N(hi)}}{\text{N(lo)}}]}\right]=\infty$$

shows that, in contradiction to our conclusions above, infinite temperature means infinite energy. In fact, the number of translational states is very large, near infinite. For a system with a Volume of about 1 cm^3 there are more translational states than molecules.

Translational states thus cannot be described by the above model. Free particles that store energy in their translational motion cannot have negative Kelvin temperature.

If we put negative temperature into the Boltzmann formula, the function value goes up exponetially. Each energy level must be populated more than its lower neighbour. Of course it is not possible to keep populating levels like this. It follows that in system that have inversely populated states, these states cannot be in thermal equilibrium with all other states of the same system. Negative absolute temperature is not an equilibrium quantity.

REFERENCES

- A Guide to Physics: Thermodynamics, Statistical Physics, and Quantum Mechanics by Gerald D. Mahan, Boris E. Nadgorny, and Max Dresden.
- DK Science Encyclopedia (Revised Edition) by Susan McKeever and Martyn Foote.
- Statistical Physics, L. D. Landau and E. M. Lifshitz
- Statistical Mechanics, S-K Ma
- Course in Thermodynamics . Revised Printing. Volume II. (Series in Thermal and Fluids Engineering) by Joseph Kestin.

13

Power Cycles with Two-Phase Media

BEHAVIOUR OF TWO-PHASE SYSTEMS

The definition of a phase, as given by SB&VW, is "a quantity of matter that is homogeneous throughout." Common examples of systems that contain more than one phase are a liquid and its vapour and a glass of ice water. A system which has three phases is a container with ice, water, and water vapour. We wish to find the relations between phases and the relations that describe the change of phase (from solid to liquid, or from liquid to vapour) of a pure substance, including the work done and the heat transfer. To start we consider a system consisting of a liquid and its vapour in equilibrium, which are enclosed in a container under a moveable piston. The system is maintained at constant temperature through contact with a heat reservoir at temperature , so there can be heat transfer to or from the system.

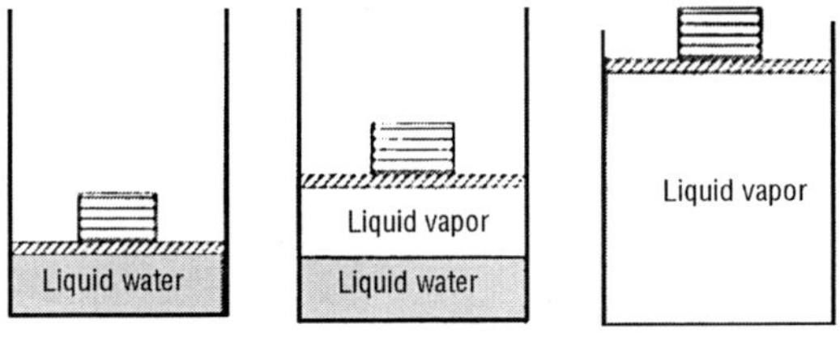

Fig . Two-phase System in Contact with Constant Temperature Heat Reservoir

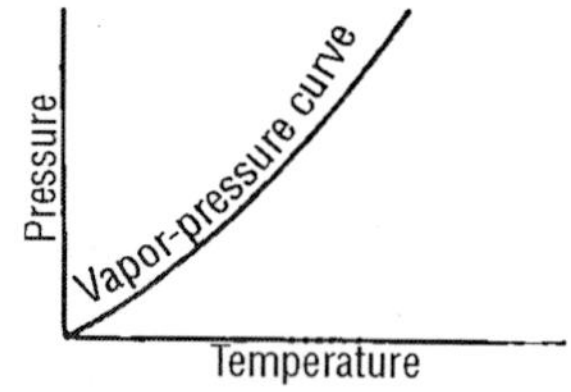

Fig. Relation for a Liquid-vapour System

For a pure substance, there is a one-to-one correspondence between the temperature at which vaporization occurs and the pressure. These values are called the saturation pressure and saturation temperature. This means there is an additional constraint for a liquid-vapour mixture, in addition to the equation of state. The consequence is that we only need to specify one variable to determine the state of the system. If we specify T then P is set. In summary, for two phases in equilibrium, $P = P(T)$. If both phases are present, any quasi-static process at constant T is also at constant P. Let us examine the pressure-volume behaviour of a liquid-vapour system at constant temperature. For a single-phase ideal gas we know that the curve would be Pv. For the two-phase system the curve looks quite different.

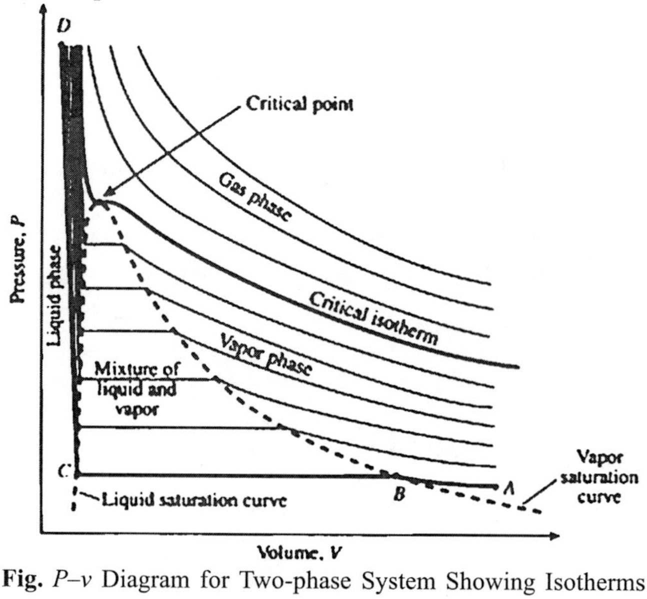

Fig. P–v Diagram for Two-phase System Showing Isotherms

Several features of the figure should be noted. First, there is a region in which liquid and vapour can coexist, bounded by the liquid saturation curve on the left and the vapour saturation curve on the right. This is roughly dome-shaped and is thus often referred to as the "vapour dome." Outside of this regime, the equilibrium state will be a single phase.

The regions of the diagram in which the system will be in the liquid and vapour phases respectively are indicated. Second is the steepness of the isotherms in the liquid phase, due to the small compressibility of most liquids.

Third, the behaviour of isotherms at temperatures below the "critical point" in the region to the right of the vapour dome approach those of an ideal gas as the pressure decreases, and the ideal gas relation is a good approximation in this region.

The behaviour shown is found for all the isotherms that go through the vapour dome. At a high enough temperature, specifically at a temperature corresponding to the pressure at the peak of the vapour dome, there is no transition from liquid to vapour and the fluid goes continuously from a liquid-like behaviour to a gas-type behaviour.

This behaviour is unfamiliar, mainly because the temperatures and pressures are not ones that we typically experience; for water the critical temperature is 374°C and the associated critical pressure is 220 atmospheres.

There is a distinct nomenclature used for systems with more than one phase. In this, the terms "vapour" and "gas" seem to be used interchangeably. In the zone where both liquid and vapour exist, there are two bounding situations.

When the last trace of vapour condenses, the state becomes saturated liquid. When the last trace of liquid evaporates the state becomes saturated vapour (or dry vapour). If we put heat into a saturated vapour it is referred to as superheated vapour. Nitrogen at room temperature and pressure (at one atmosphere the vaporization temperature of nitrogen is 77 K) is a superheated vapour.

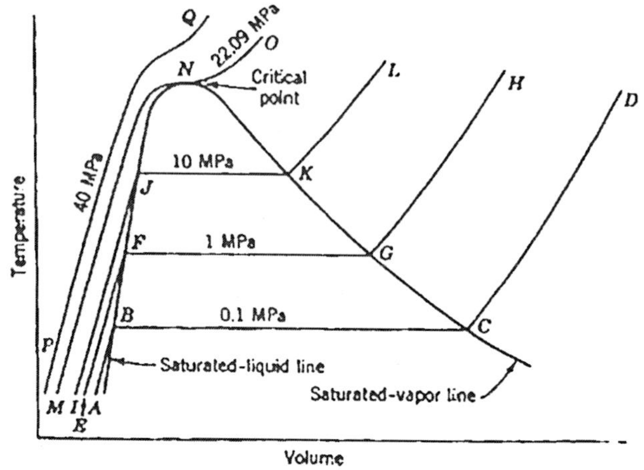

Fig. Constant Pressure Curves in *T-v* Coordinates Showing Vapour Dome

Lines of constant pressure in temperature-volume coordinates. Inside the vapour dome the constant pressure lines are also lines of constant temperature. It is useful to describe the situations encountered as we decrease the pressure or equivalently increase the specific volume, starting from a high pressure, low specific volume state.

The behaviour in this region is liquid-like with very little compressibility. As the pressure is decreased, the volume changes little until the boundary of the vapour dome is reached. Once this occurs, however, the pressure is fixed because the temperature is constant. As the piston is withdrawn, the specific volume increases through more liquid evaporating and more vapour being produced. During this process, since the expansion is isothermal (we specified that it was), heat is transferred to the system.

The specific volume will increase at constant pressure until the right hand boundary of the vapour dome is reached. At this point, all the liquid will have been transformed into vapour and the system again behaves as a single-phase fluid. For water at temperatures near room temperature, the behaviour would be essentially that of a perfect gas in this region. To the right of the vapour dome, as mentioned above, the behaviour is qualitatively like that of a perfect gas.

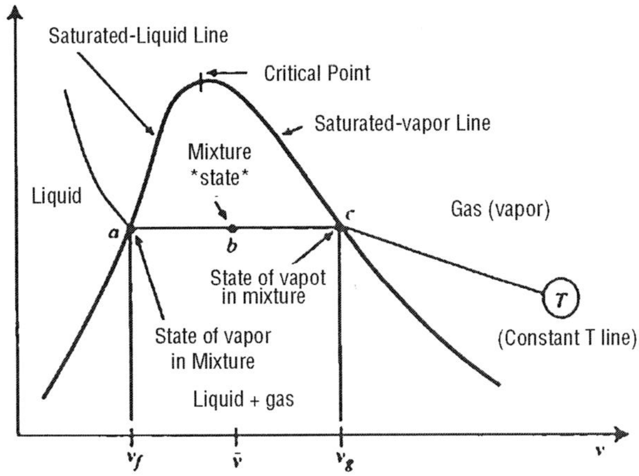

Fig. Specific Volumes at Constant Temperature and States within the Vapour Dome in a Liquid-vapour System

We define notation to be used in what follows. The states *a* and *c* denote the conditions at which all the fluid is in the liquid state and the gaseous state respectively.

The specific volumes corresponding to these states are

$v_f \equiv$ specific volume of liquid phase,.

$v_g \equiv$ specific volume of gas phase.

For conditions corresponding to specific volumes between these two values, i.e., for state *b*, the system would exist with part of the mass in a liquid state and part of the mass in a gaseous (vapour) state. The average specific volume for this condition is

$\overline{v} \equiv$ average specific volumeof two-phase system

We can relate the average specific volume to the specific volumes for liquid and vapour and the mass that exists in the two phases as follows. The total mass of the system is given by

total mass $= m =$ liquid mass + vapour mass $= m_f + m_g$.

The volume of the system is

$$\text{Volume of liquid} = V_f = m_f v_f$$
$$\text{Volume of liquid} = V_g = m_g v_g$$
$$\text{Total volume} = V = m_f v_f + m_g v_g.$$

The average specific volume, $\overline{v}$, is the ratio of the total volume to the total mass of the system

$$\bar{v} = \frac{m_f v_f + m_g v_g}{m_f + m_g} = \text{average specific volume.}$$

The fraction of the total mass in the vapour phase is called quality, and denoted by X:

$$X = \frac{m_g}{m_f + m_g} = \text{quanlity of a liquid-vapour system}$$

In terms of the quality and specific volumes, the average specific volume can be expressed as $\bar{v} = Xv_g + (1 - X)\, v_f$.

$$ab = \bar{v} - v_f \; a_c = v_g - v_f$$

In reference to Figure,

$$\frac{ab}{ac} = \frac{\bar{v} - v_f}{v_g - v_f} = X = \text{quality.}$$

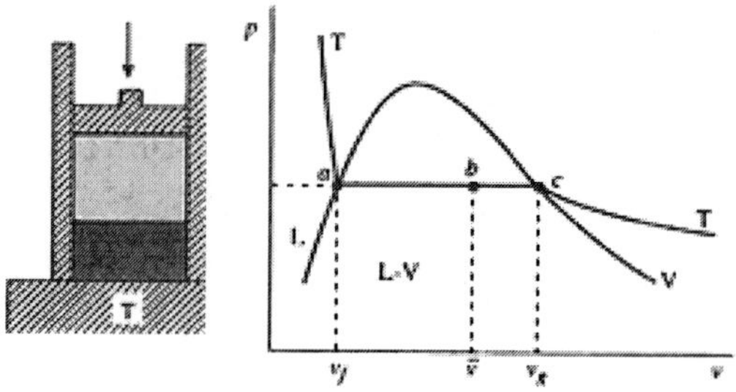

Fig. Liquid Vapour Equilibrium in a Two-phase Medium

WORK AND HEAT TRANSFER WITH TWO-PHASE MEDIA

We examine the work and heat transfer in quasi-static processes with two-phase systems. For definiteness, consider the system to be a liquid-vapour mixture in a container whose volume can be varied through movement of a piston. The system is kept at constant temperature through contact with a heat reservoir at temperature T.

The pressure is thus also constant, but the volume, V, can change. For a fixed mass, the volume is proportional to the specific volume so that point in Figure must move to the left or the right as V changes. This implies that the amount of mass in each of the two phases, and hence the quality, also changes because mass is transferred from one phase to the other. We wish to find the heat and work transfer associated with the change in mass in each phase. The change in volume can be related to the changes in mass in the two phases as,

$$dV = v_g dm_g + v_f dm_f.$$

The system mass is constant ($m = m_f + m_g$ = constant) so that for any changes

$$dm = 0 = dm_f + dm_g.$$

We can define the quantity

$dm_{fg} = dm_g = - dm_f$ = mass transferred from liquid to vapour.

In terms of dm_{fg} the volume change of the system is

$$dV = (v_g - v_f)\, dm_{fg}.$$

The work done is given by

$$dW = PdV = P(v_g - v_f)\, dm_{fg}.$$

The change in internal energy, ΔU, can be found as follows. The internal energy of the system can be expressed in terms of the mass in each phase and the specific internal energy (internal energy per unit mass, u) of the phase as,

$$U = u_f m_f + u_g m_g$$

$$dU = u_f dm_f + u_g dm_g = (u_g - u_f)\, dm_{fg}.$$

Note that the specific internal energy of the two-phase system can be expressed in a similar way as the specific volume in terms of the quality and the specific internal energy of each phase:

$$u = Xu_g + (1 - X)\, u_f.$$

Writing the first law for this process:

$$\begin{aligned} dQ &= dU + dW \\ &= (u_g - u_f)dm_{fg} + P(v_g - v_f)dm_{fg} \\ &= [(u_g + Pv_g) - (u_f + Pv_f)]dm_{fg} \\ &= (h_g - h_f)dm_{fg}. \end{aligned}$$

The heat needed for the transfer of mass is proportional to the difference in specific enthalpy between vapour and liquid. The pressure and temperature are constant, so that the specific internal energy and the specific enthalpy for the liquid phase and the gas phase are also constant. For a finite change in mass from liquid to vapour, m_{fg}, therefore, the quantity of heat needed is

$$Q = (h_g - h_f) = \Delta H \text{ (enthalpy change)}$$

The heat needed per unit mass, q, for transformation between the two phases is

$$q = \frac{Q}{m_{fg}} = (h_g - h_f) = h_{fg}.$$

The notationm h_{fg} refers to the specific enthalpy change between the liquid state and the vapour state. The expression for the amount of heat needed, q, is a particular case of the general result that in any reversible process at constant pressure, the heat flowing into, or out of, the system is equal to the enthalpy change. Heat is absorbed if the change is from solid to liquid (heat of fusion), liquid to vapour (heat of vaporization), or solid to vapour (heat of sublimation).

A numerical example is furnished by the vaporization of water at 100°C:

- How much heat is needed per unit mass of fluid vaporized?
- How much work is done per unit mass of fluid vaporized?
- What is the change in internal energy per unit mass of fluid vaporized?

In addressing these questions, we make use of the fact that problems involving heat and work exchanges in two-phase media are important enough that the values of the specific thermodynamic properties that characterize these transformations have been computed for many different working fluids. From these, for water:

- At 100°C, the vapour pressure is 0.1013 MPa.
- The specific enthalpy of the vapour, h_g, is 2676 kJ/kg and the specific enthalpy of the liquid, h_f, is 419 kJ/kg.
- The difference in enthalpy between liquid and vapour, h_{fg}, occurs often enough so that it is tabulated also. This is 2257 kJ/kg.
- The specific volume of the vapour is 1.6729 m^3/kgand the specific volume of the liquid is 0.001044 m^3/kg.

The heat input to the system is the change in enthalpy between liquid and vapour, h_{fg}, and is equal to 2.257×106 J/kg.

The work done is $P(v_g - v_f)$which has a value of

$$P(v_g - v_f) = 0.1013 \times 10^6 \text{ Pa} \times [1.629 \text{ m}^3/\text{kg}]$$
$$= 0.169 \times 10^6 \text{ J/kg}.$$

The change in internal energy per unit mass (u_{fg}) can be found from $\Delta u = q - w$ or from the tabulated values as $2.088{\times}10^6$J/kg. This is much larger than the work done. Most of the heat input is used to change the internal energy rather than appearing as work.

THE CARNOT CYCLE AS A TWO-PHASE POWER CYCLE

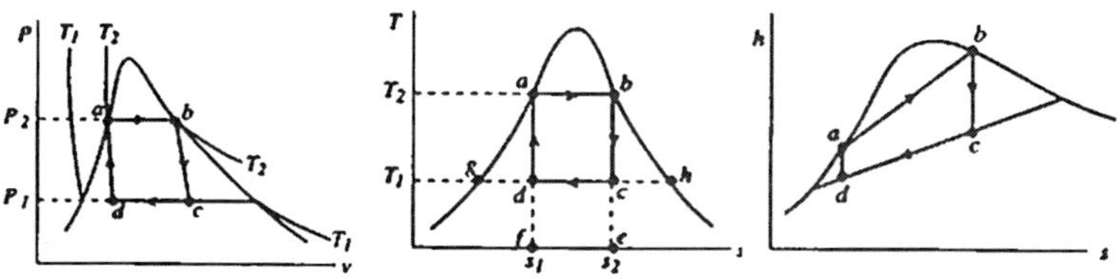

[cycle in P-v coordinates] [cycle in T-s coordinates] [cycle in h-s coordinates]

Fig. Carnot cycle with two-phase medium

A Carnot cycle that uses a two-phase fluid as the working medium. Figure gives the cycle in *P-v* coordinates, Figure in *T-s* coordinates, and figure in *h-s* coordinates. The boundary of the region in which there is liquid and vapour both present (the vapour dome) is also indicated. Note that the form of the cycle is different in the *T-s* and *h-s* representation; it is only for a perfect gas with constant specific heats that cycles in the two coordinate representations have the same shapes.

The processes in the cycle are as follows:

- Start at state *a* with saturated liquid (all of mass in liquid condition). Carry out a reversible isothermal expansion to $b(a \rightarrow b)$ until all the liquid is vaporized. During this process a quantity of heat *qH* per unit mass is received from the heat source at temperature T_2.
- Reversible adiabatic (i.e., isentropic) expansion ($b \rightarrow c$) lowers the temperature to T_1. Generally state will be in the region where there is both liquid and vapour.
- Isothermal compression ($c \rightarrow d$) at to T_1 state *d*. During

this compression, heat qL per unit mass is rejected to the source at T_1.

- Reversible adiabatic (i.e., isentropic) compression ($d \to a$) in which the vapour condenses to liquid and the state returns to a.

In the T-s diagram the heat received, qH, is *abef* and the heat rejected, qL, is *dcef*. The net work is represented by *abcd*. The thermal efficiency is given by

$$\eta = \frac{w_{\text{net}}}{qH} = \frac{\text{Area abcd}}{\text{Area abef}} = 1 - \frac{T_1}{T_2}.$$

In the g-s diagram, the isentropic processes are vertical lines as in the T-s diagram. The isotherms in the former, however, are not horizontal as they are in the latter. To see their shape we note that for these two-phase processes the isotherms are also lines of constant pressure (isobars), since $P = P(T)$. The combined first and second law is

$$Tds = dh - \frac{dp}{\rho}.$$

For a constant pressure reversible process, $dq_{\text{rev}} = Tds = dh$. The slope of a constant pressure line in h-s coordinates is thus,

$\left(\frac{\partial h}{\partial s}\right)_P = T = \text{constant}$; slope of constant pressure line for two-phase medium.

The heat received and rejected per unit mass is given in terms of the enthalpy at the different states as

$qH = h_b - h_a$

$q_L = h_d - h_c$. (In accord with our convention this is less than zero.)

The thermal efficiency is

$$\eta = \frac{w_{\text{net}}}{q_H} = \frac{q_H + q_L}{q_H} = \frac{(h_b - h_a) + (h_d - h_c)}{(h_b - h_a)},$$

or, in terms of the work done during the isentropic compression and expansion processes, which correspond to the shaft work done on the fluid and received by the fluid,

$$\eta = \frac{(h_b - h_c) + (h_a - h_d)}{(h_b - h_a)},$$

Carnot Steam Cycle

- Heat source temperature = 300°C
- Heat sink temperature = 20°C

What is the (i) thermal efficiency and (ii) ratio of turbine work to compression (pump) work if

- All processes are reversible?
- The turbine and the pump have adiabatic efficiencies of 0.8?

Neglect the changes in kinetic energy at inlet and outlet of the turbine and pump.

- For the reversible cycle,

$$\eta_{\text{thermal}} = \eta_{\text{Carnot}} = 1 - \frac{T_1}{T_2}$$

$$= 1 - \frac{293\ K}{573\ K} = 0.489.$$

- To find the work in the pump (compression process) or in the turbine, we need to find the enthalpy changes between states b and c, Δh_{bc}, and the change between a and d, Δh_{ad}. To obtain these the approach is to use the fact that s = constant during the expansion to find the quality at state and then, knowing the quality, calculate the enthalpy as $h = Xh_g + (1 - X)\ h_f$. We know the conditions at state b, where the fluid is all vapour, i.e., we know T_b, h_b, s_b:
  $h_b = h_{\text{vapor}}$ (300°C) $= h_g$ (300°C) = 2749 kJ/kg
  $s_b = s_{\text{vapor}}$ (300°C) $= s_g$ (300°C) = 5.7045 kJ/kg-K.
  $s_b = s_c$ in the isentropic expansion process.
- We now need to find the quality at state c, X_c. Using the definition of quality, and noting that $s_c = X_c s_g + (1 - X_c)s_f$, we obtain,
- $$X_c = \frac{s_c - s_f(T_c)}{s_g(T_c) - s_f(T_c)} = \frac{s_c - s_f(T_c)}{s_{fg}(T_c)}.$$
- The quantity s_c is the mass-weighted entropy at state c, which is at temperature T_c.

- The quantity $s_f(T_c)$is the entropy of the liquid at temperature T_c.
- The quantity $s_f(T_c)$ is the entropy of the gas (vapour) at temperature T_c.
- The quantity $\Delta s_{fg}\,(T_c) = \Delta s_{\text{liquid}\rightarrow\text{gas}}$ at T_c.

We know:

$$s_c = s_b = 5.7045 \text{ kJ/kg-K},$$
$$s_{fg} = 8.3706 \text{ kJ/kg-K},$$

The quality at state c is thus,

$$X_c = \frac{5.7045 - 0.2966}{8.3706} = 0.646.$$

The enthalpy at state is,

$$h_c = X_c h_g + (1 - X_c)\, h_f \text{ at } T_c.$$

Substituting the values,

$$h_c = 0.646 \times 2538.1 \text{ kJ/kg} + 0.354 \times 83.96 \text{ kJ/kg}$$
$$= 1669.4 \text{ kJ/kg}.$$

The turbine work/unit mass is the difference between the enthalpy at state b and state c,

$$h_b - h_c = w_{\text{turbine}} = 2749 - 1669.4 = 1079.6 \text{ kJ/kg}.$$

We can apply a similar process to find the conditions at state d:

$$X_d = \frac{s_d - s_f(T_d)}{s_g(T_d) - s_f(T_d)} = \frac{s_c - s_f(T_d)}{s_{fg}(T_d)}.$$

We have given that $T_c = T_d$. Also $s_d = s_a = s_f$ at 300°C. The quality at state d is

$$X_d = \frac{3.253 - 0.2966}{8.3706} = 0.353 < X_c.$$

The enthalpy at state d is

$$h_d = X_d h_g + (1 - X_d) h_f = 0.353 \times 2538.1 \text{ kJ/kg} + 0.647 \times 83.96 \text{ kJ/kg}$$
$$= 950.8 \text{ kJ/kg}.$$

The work of compression (pump work) is $\Delta h_{ad} = h_a - h_d$. Substituting the numerical values,

$$\Delta h_{ad} = 1344 - 950.8 = 393.3 \text{ kJ/kg}.$$

The ratio of turbine work to compression work (pump work) is

$$\frac{w_{\text{turbine}}}{w_{\text{compression}}} = 2.75.$$

We can check the efficiency by computing the ratio of net work ($w_{\text{net}} = w_{\text{turbine}} - w_{\text{compression}}$) to the heat input ($T_a s_{fg}$). Doing this gives, not surprisingly, the same value as the Carnot equation.

- For a cycle with adiabatic efficiencies of pump and turbine both equal to 0.8 (non-ideal components), the efficiency and work ratio can be found as follows.

We can find the turbine work using the definition of turbine and compressor adiabatic efficiencies. The relation between the enthalpy changes is

$w_{\text{turbine}} = h_b - h_c' = \eta_{\text{turbine}}(h_b - h_c)$ = actual turbine work received. Substituting the numerical values, the turbine work per unit mass is 863.7 kJ/k.

For the compression process, we use the definition of compressor (or pump) adiabatic efficiency:

$$w_{\text{compression}} = h_a' - h_d = \frac{1}{\eta_{\text{compression}}}(h_a - h_d)$$

= actual work to achieve given pressure difference

= 491.6 kJ/kg.

The value of the enthalpy at state is a' 1442.4 kJ/kg. The thermal efficiency is given by

$$= \frac{w_{\text{net}}}{\text{heat input}} = \frac{w_{\text{turbine}} - w_{\text{compression}}}{\text{heat input}}$$

$$= \frac{(h_b - h_{c'}) - (h_{a'} - h_d)}{(h_b - h_{a'})}.$$

Substituting the numerical values, we obtain for the thermal efficiency with non-ideal components, $\eta_{\text{thermal}} = 0.285$

A question arises as to whether the Carnot cycle can be practically applied for power generation. The heat absorbed and the heat rejected both take placc at constant temperature and pressure within the two-phase region. These can be closely approximated by a boiler for the heat addition process and a condenser for the heat rejection. Further, an efficient turbine can produce a reasonable approach to reversible adiabatic expansion, because the steam is expanded with only small losses. The difficulty occurs in the compression part of the cycle.

If compression is carried out slowly, there is equilibrium between the liquid and the vapour, but the rate of power generation may be lower than desired and there can be appreciable heat transfer to the surroundings. Rapid compression will result in the two phases coming to very different temperatures (the liquid temperature rises very little during the compression whereas the vapour phase temperature changes considerably). Equilibrium between the two phases cannot be maintained and the approximation of reversibility is not reasonable.

Another circumstance is that in a Carnot cycle all the heat is added at the same temperature. For high efficiency we need to do this at a higher temperature than the critical point, so that the heat addition no longer takes place in the two-phase region. Isothermal heat addition under this circumstance is difficult to accomplish. Also, if the heat source and the cycle are considered together, the products of combustion which provide the heat can be cooled only to the highest temperature of the cycle.

The source will thus be at varying temperature while the system requires constant temperature heat addition, so there will be irreversible heat transfer. In summary, the practical application of the Carnot cycle is limited because of the inefficient compression process, the low work per cycle, the upper limit on temperature for operation in the two-phase flow regime, and the irreversibility in the heat transfer from the heat source. We examine the Rankine cycle, which is much more compatible with the characteristics of two-phase media and available machinery for carrying out the processes.

THE CLAUSIUS-CLAPEYRON EQUATION

Until now we have only considered ideal gases and we would like to show that the properties u, h, s, etc. are true state variables and that the 1st and 2nd laws of thermodynamics hold when the working medium is not an ideal gas (i.e. a two-phase medium). An elegant way to do this is to consider a Carnot cycle for a two-phase medium. To state the fact that all Carnot engines operated between two given temperatures have the same efficiency is one way of stating the 2nd law of thermodynamics. The working fluid need not be an ideal gas and may be a two-phase medium changing phases.

The idea is to run a Carnot engine between temperatures T and $T - dT$ for a two-phase medium and to let it undergo a change in phase. We can then derive an important relation known as the Clausius-Clapeyron equation, which gives the slope of the vapour pressure curve. We could then measure the vapour pressure curve for various substances and compare the measured slope to the Clausius-Clapeyron equation. This can then be viewed as an experimental proof of the general validity of the 1st and 2nd laws of thermodynamics!

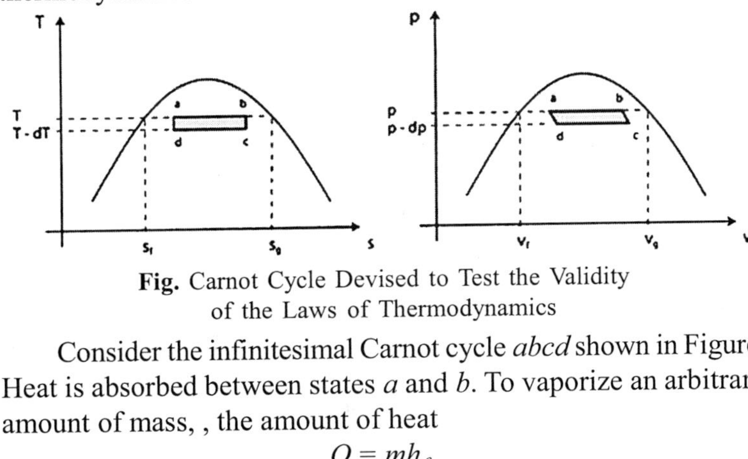

Fig. Carnot Cycle Devised to Test the Validity of the Laws of Thermodynamics

Consider the infinitesimal Carnot cycle *abcd* shown in Figure. Heat is absorbed between states *a* and *b*. To vaporize an arbitrary amount of mass, , the amount of heat

$$Q = mh_{fg}$$

must be supplied to the system. From the 1st and 2nd laws of thermodynamics the thermal efficiency for a Carnot cycle can be written as

$$\eta = \frac{W}{Q} = \frac{Q - Q_R}{Q} = \frac{T - T_R}{T}.$$

Hence, for the infinitesimal cycle considered above,

$$\frac{dW}{Q} = \frac{T - (T - dT)}{T} = \frac{dT}{T}.$$

The work along *bc* and *da* nearly cancel such that the net work is the difference between the work along *ab* and *cd*, and *dW* can be viewed as the area enclosed by the rectangle *abcd*:

$$dW = \text{pm}\,(v_g - v_f) - (p - d_p)\, m\,(v_g - v_f)$$
$$= m(v_g - v_f)\, d_p.$$

Substituting Equations we get

$$\frac{m(v_g - v_f)dp}{mh_{fg}} = \frac{dT}{T}.$$

Rearranging terms yields the Clausius-Clapeyron equation, which defines the slope of the vapour pressure curve:

$$\frac{dp}{dT} = \frac{h_{fg}}{T(v_g - v_f)}.$$

The beauty is that we have found a general relation between experimentally measurable quantities from first principles (1st and 2nd laws of thermodynamics).

In order to plot the Clausius-Clapeyron relation and to compare it against experimentally measured vapour pressure curves, we need to integrate Equation. To do so, the heat of vaporization and the specific volumes must be known functions of temperature. This is an important problem in physical chemistry but we shall not pursue it further here except to mention that if

- Variations in heat of vaporization can be neglected,
- The vapour phase is assumed to be an ideal gas, and
- The specific volume of the liquid is small compared to that of the vapour phase,

$$h_{fg} \approx \text{const}, \quad v_f << v_g \approx \frac{RT}{p},$$

the integration can be readily carried out. Making these approximations, the Clausius-Clapeyron equation becomes

$$\frac{dp}{dT} \approx \frac{h_{fg}P}{T^2R} \quad \longrightarrow \quad \frac{dp}{p} \approx \frac{h_{fg}}{R}\frac{dT}{T^2}.$$

Carrying out the integration, the resulting expression is

$$\ln p = \frac{h_{fg}}{R}\frac{1}{T} + \ln C.$$

Note that the vapour pressure curves are straight lines if ln p is plotted versus $1/T$ and that the slope of the curves is $-h_{fg}/R$, directly related to the heat of vaporization.

RANKINE POWER CYCLES

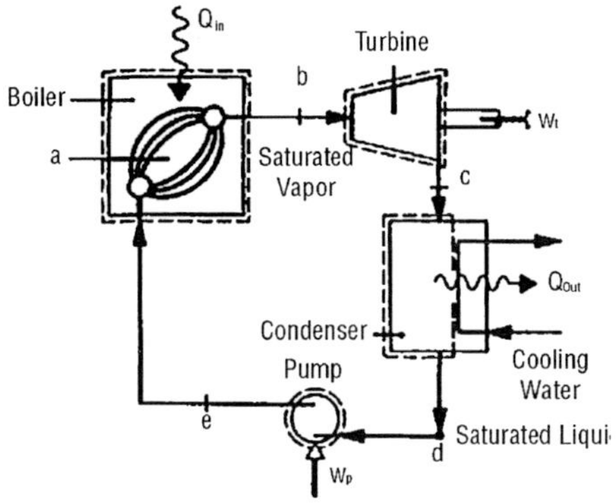

Fig. Rankine Power Cycle with Two-phase Working Fluid

A schematic of the components of a Rankine cycle is shown in Figure. The cycle is shown on *P-v*, *T-s*, and *h-s* coordinates in Figure. The processes in the Rankine cycle are as follows:

- $d \to e$: Cold liquid at initial temperature T_1 is pressurized reversibly to a high pressure by a pump. In this process, the volume changes slightly.
- $e \to a$: Reversible constant pressure heating in a boiler to temperature T_2.
- $a \to b$: Heat added at constant temperature T_2 (constant pressure), with transition of liquid to vapour.
- $b \to c$: Isentropic expansion through a turbine. The quality decreases from unity at point b to $X_c < 1$.
- $c \to d$: Liquid-vapour mixture condensed at temperature T_1 by extracting heat.

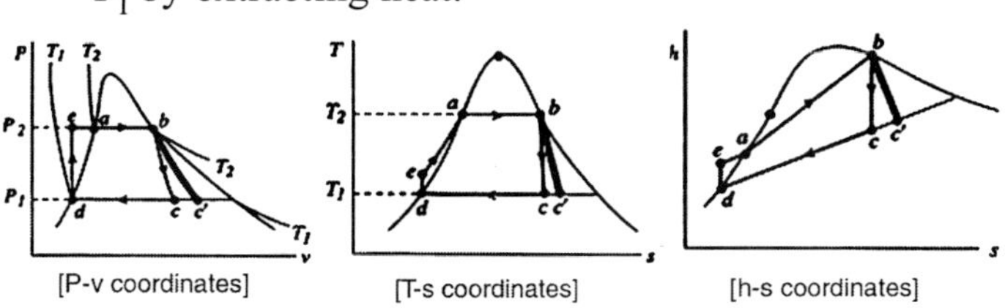

Fig. Rankine Cycle Diagram.

In the Rankine cycle, the mean temperature at which heat is supplied is less than the maximum temperature, T_2, so that the efficiency is less than that of a Carnot cycle working between the same maximum and minimum temperatures. The heat absorption takes place at constant pressure over *eab*, but only the part *ab* is isothermal. The heat rejected occurs over *cd*; this is at both constant temperature and pressure.

To examine the efficiency of the Rankine cycle, we define a mean effective temperature, T_m, in terms of the heat exchanged and the entropy differences:

$$q_H = T_{m2}\Delta s_2.$$

The thermal efficiency of the cycle is

$$\eta_{\text{thermal}} = \frac{T_{m2}(s_b - s_c) - T_{m1}(s_c - s_d)}{T_{m2}\,(s_b - s_c)}$$

The compression and expansion processes are isentropic, so the entropy differences are related by

$$s_b - s_c = s_c - s_d.$$

The thermal efficiency can be written in terms of the mean effective temperatures as

$$\eta_{\text{thermal}} = 1 - \frac{T_{m1}}{T_{m2}}.$$

For the Rankine cycle, $Tm_1 \approx T_1$ $T_{m2} < T_2$. From this equation we see not only the reason that the cycle efficiency is less than that of a Carnot cycle, but the direction to move in terms of cycle design (increased T_{m2}) if we wish to increase the efficiency.

There are several features that should be noted about Figure and the Rankine cycle in general:

- The *T-s* and the *h-s* diagrams are not similar in shape, as they were with the perfect gas with constant specific heats. The slope of a constant pressure reversible heat addition line is,

$$\left(\frac{\partial h}{\partial s}\right)_P = T.$$

In the two-phase region, constant pressure means also constant temperature, so the slope of the constant pressure heat addition line is constant and the line is straight.

- The effect of irreversibilities is represented by the dashed line from b to c'. Irreversible behaviour during the expansion results in a value of entropy $s_c{}'$ at the end state of the c' expansion that is higher than s_c. The enthalpy at the end of the expansion (the turbine exit) is thus higher for the irreversible process than for the reversible process, and, as seen for the Brayton cycle, the turbine work is thus lower in the irreversible case.
- The Rankine cycle is less efficient than the Carnot cycle for given maximum and minimum temperatures, but, as said earlier, it is more effective as a practical power production device.

EFFECT OF DESIGN PARAMETERS ON RANKINE CYCLES

The basic Rankine cycle can be enhanced through processes such as superheating and reheat. Diagrams for a Rankine cycle with superheating are given in Figure. The heat addition is continued past the point of vapour saturation, in other words the vapour is heated so that its temperature is higher than the saturation temperature associated with $P_a (= P_b = P_c = P_d)$. This does several things. First, it increases the mean temperature at which heat is added, T_{m2}, thus increasing the efficiency of the cycle.

Second is that the quality of the two-phase mixture during the expansion is higher with superheating, so that there is less moisture content in the mixture as it flows through the turbine. This is an advantage in terms of decreasing the mechanical deterioration of the blading.

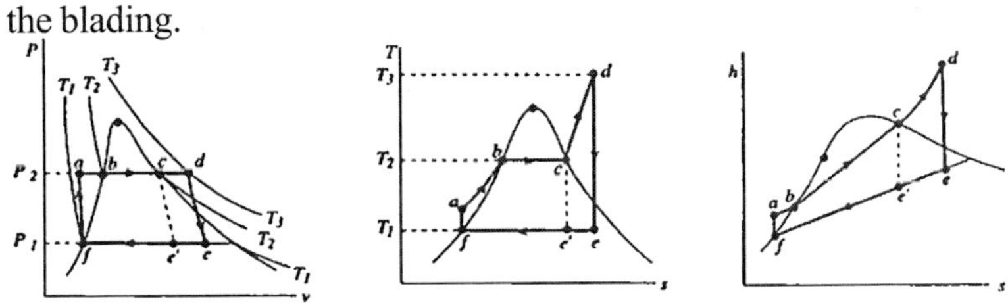

[P - v coordinates] [T - s coordinates] [h - s coordinates]

Fig. Rankine Cycle with Superheating

The heat exchanges in the superheated cycle are:

- Along *abcd*, which is a constant pressure (isobaric) process: $q_2 = h_d = h_a$.

- Along $e_f: q_1 = h_f - h_c$, (< 0).

The thermal efficiency of the ideal Rankine cycle with superheating is

$$\eta_{\text{thermal}} = \frac{h_d - h_a - (h_c - h_f)}{h_d - h_a}.$$

This can be expressed explicitly in terms of turbine work and compression (pump) work as

$$\eta_{\text{thermal}} = \frac{h_d - h_c - (h_a - h_f)}{h_d - h_a}.$$

Compared to the basic cycle, superheating has increased the turbine work, increased the mean temperature at which heat is received, T_{m2}, and increased the cycle efficiency.

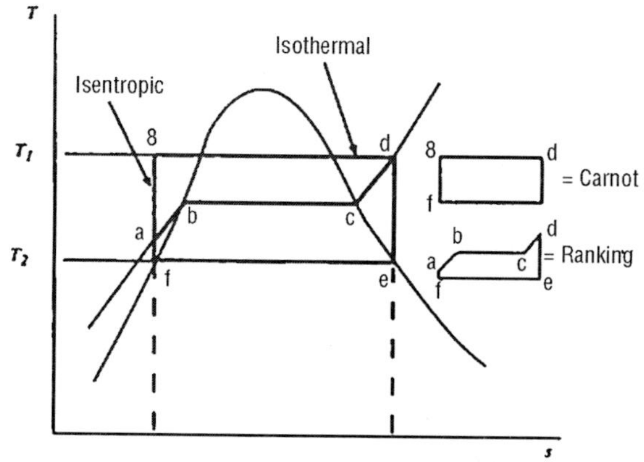

Fig. Comparison of Rankine Cycle with Superheating and Carnot Cycle

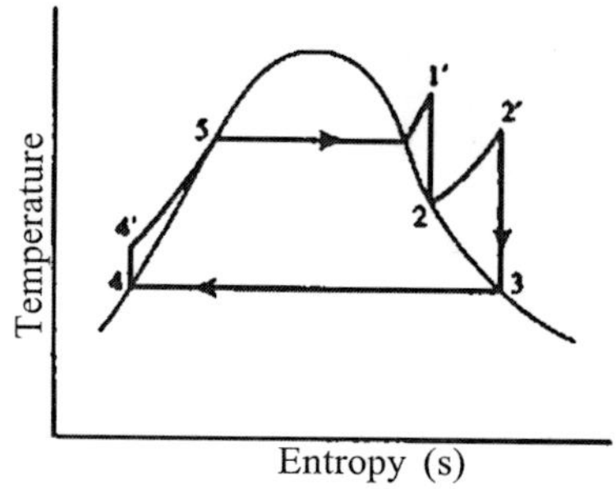

Fig. Rankine Cycle with Reheat

A comparison of the Carnot cycle and the Rankine cycle with superheating is given in Figure. The maximum and minimum temperatures are the same, but the average temperature at which heat is absorbed is lower for the Rankine cycle. To alleviate the problem of having moisture in the turbine, one can heat again after an initial expansion in a turbine, which gives a schematic of a Rankine cycle for space power application. This process is known as reheat.

The main practical advantage of reheat (and of superheating) is the decrease in moisture content in the turbine because most of the heat addition in the cycle occurs in the vaporization part of the heat addition process.

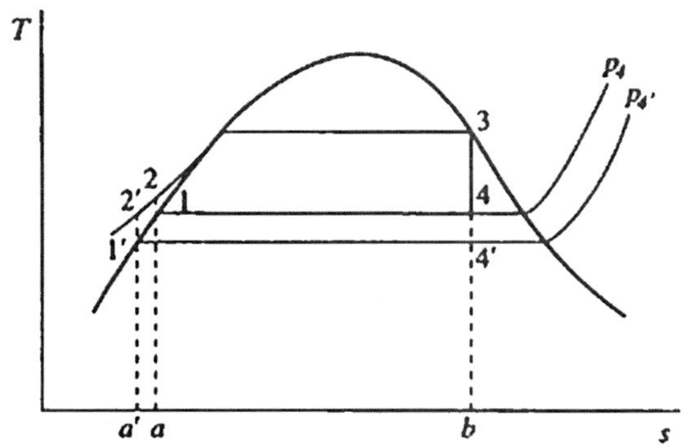

Fig. Effect of Exit Pressure on Rankine Cycle Efficiency

We can also examine the effect of variations in design parameters on the Rankine cycle. Consider first the changes in cycle output due to a decrease in exit pressure. In terms of the cycle, the exit pressure would be decreased from P_4 to $P_{4'}$. The original cycle is 1 – 2 – 3 – 4 – 1, and the modified cycle is 1′ – 2′ – 3′ – 4′ – 1′.

The consequences are that the cycle work, which is the integral of *Tds* around the cycle, is increased. In addition, as drawn, although the levels of the mean temperature at which the heat is absorbed and rejected both decrease, the largest change is the mean temperature of the heat rejection, so that the thermal efficiency increases.

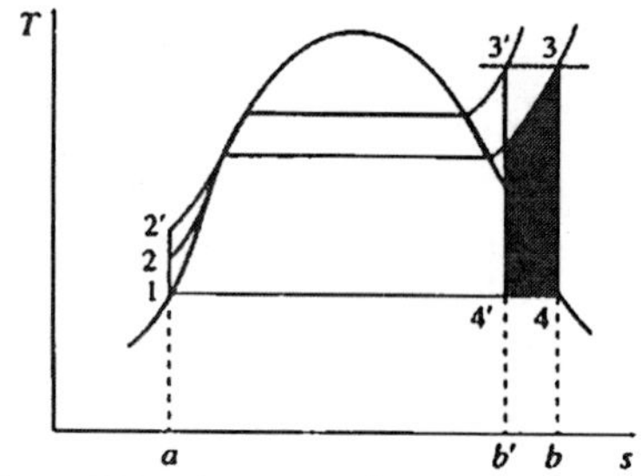

Fig. Effect of Maximum Boiler Pressure on Rankine Cycle Efficiency

Another design parameter is the maximum cycle pressure. As comparison of two cycles with different maximum pressure but the same maximum temperature, which is set by material properties. The average temperature at which the heat is supplied for the cycle with a higher maximum pressure is increased over the original cycle, so that the efficiency increases.

COMBINED CYCLES IN GAS TURBINE FOR POWER PRODUCTION

The turbine entry temperature in a gas turbine (Brayton) cycle is considerably higher than the peak steam temperature. Depending on the compression ratio of the gas turbine, the turbine exhaust temperature may be high enough to permit efficient generation of steam using the "waste heat" from the gas turbine. A configuration such as this is known as a gas turbine-steam combined cycle power plant.

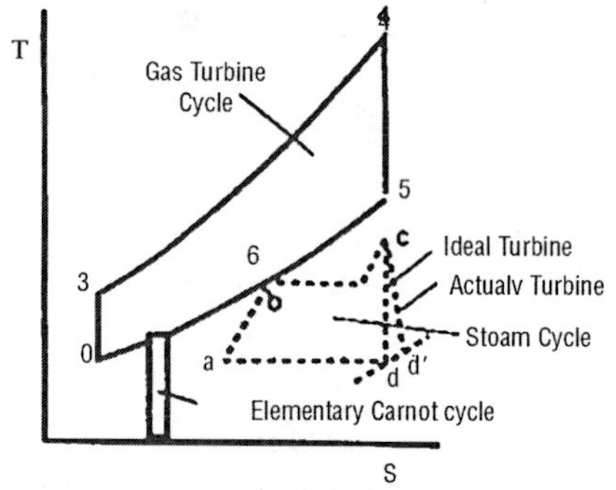

Fig. Gas Turbine-steam Combined Cycle

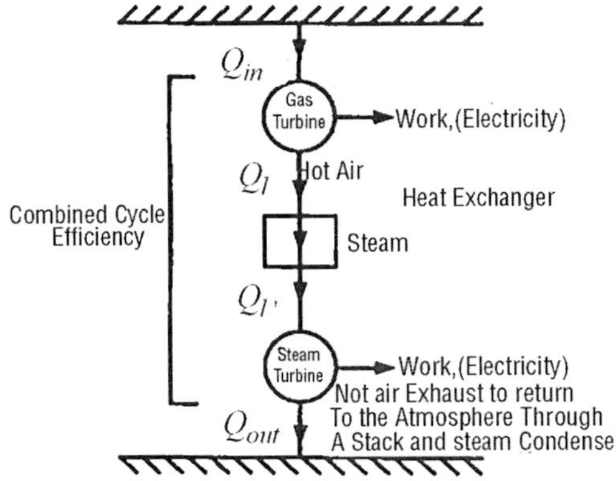

Fig. Schematic of Combined Cycle Using Gas Turbine (Brayton Cycle) and Steam Turbine (Rankine Cycle)

The heat input to the combined cycle is the same as that for the gas turbine, but the work output is larger (by the work of the Rankine cycle steam turbine). A schematic of the overall heat engine, which can be thought of as composed of an upper and a lower heat engine in series. The upper engine is the gas turbine (Brayton cycle) which expels heat to the lower engine, the steam turbine (Rankine cycle).

The overall efficiency of the combined cycle can be derived as follows. We denote the heat received by the gas turbine as Q_{in} and the heat rejected to the atmosphere as Q_{out}. The heat out of the gas turbine is denoted as Q_1. The hot exhaust gases from the gas turbine pass through a heat exchanger where they are used as the heat source for the two-phase Rankine cycle, so that $Q_1$is also the heat input to the steam cycle. The overall combined cycle efficiency is

$$\eta CC = \frac{W}{Q_{in}} = \frac{W_B + W_B}{Q_{in}},$$

where the subscripts refer to combined cycle (CC), Brayton cycle (B) and Rankine cycle (R) respectively.

From the first law, the overall efficiency can be expressed in terms of the heat inputs and heat rejections of the two cycles as (using the quantity $|Q_1|$ to denote the magnitude of the heat transferred):

$$\eta CC = \frac{Q_{in} - |Q_1| + (|Q_1| - Q_{out})}{Q_{in}} = \left[1 - \frac{|Q_1|}{Q_{in}}\right] + \left[1 - \frac{Q_{out}}{|Q_1|}\right]\left(\frac{|Q_1|}{Q_{in}}\right).$$

The first square bracket term on the right hand side is the Brayton cycle efficiency, η_B, the second is the Rankine cycle efficiency, ηR, and the term in parentheses is $(1-\eta_B)$. The combined cycle efficiency can thus be written as $\eta_{CC} = \eta_B + \eta_{\underline{R}} - \eta_B\eta_R$; Combined cycle efficiency.

Equation gives insight into why combined cycles are so successful. Suppose that the gas turbine cycle has an efficiency of 40%, which is a representative value for current Brayton cycle gas turbines, and the Rankine cycle has an efficiency of 30%. The combined cycle efficiency would be 58%, which is a very large increase over either of the two simple cycles. Some representative efficiencies and power outputs for different cycles.

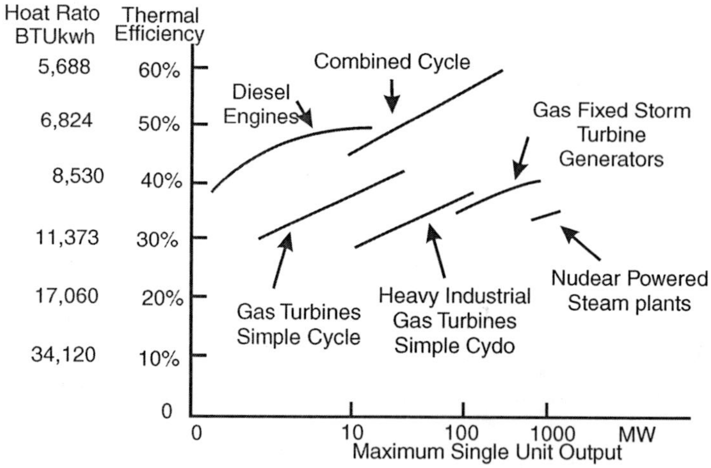

Fig. Comparison of Efficiency and Power Output of Various Power Products

Some Overall Comments on Thermodynamic Cycles

- There are many different power and propulsion cycles, and we have only looked at a few of these. Many other cycles have been devised in the search for ways to increase efficiency and power in practical devices.

- We can view a given cycle in terms of elementary Carnot cycles. This shows that the efficiency of any other cycle operating between two given temperatures will be less than that of a Carnot cycle.
- If we view the thermal efficiency as

$$\eta_{\text{thermal}} = 1 - \frac{(T_{\text{heat rejected}})\,\text{Average}}{(T_{\text{heat absorbed}})\,\text{Average}},$$

  this means that we should accept heat at a high temperature and reject it at a low temperature for high efficiency. This objective must be tempered by considerations of practical application.
- The cycle diagrams in *T-s* and *h-s* coordinates will only be similar if the working medium is an ideal gas. For other media (, a two-phase mixture) they will look different.
- Combined cycles make use of the rejected heat from a "topping" cycle as heat source for a "bottoming" cycle. The overall efficiency is higher than the efficiency of either cycle.

REFERENCES

- Heat and Thermodynamics: A Historical Perspective (Greenwood Guides to Great Ideas in Science) by Christopher J. T. Lewis
- Equilibrium Statistical Physics, M. Plische and B. Bergersen
- A Modern Course in Statistical Physics, L. E. Reichl
- Thermodynamics by J.P. Holman.
- Statistical mechanics, R. P. Feynman, W. A. Benjamin.

Bibliography

- Statistical Mechanics, *K. Huang,*
- Statistical Physics, *L. D. Landau* and *E. M. Lifshitz*
- Modern Thermodynamics, D. Kondepundi and I. Prigogine
- Statistical Mechanics, *S-K Ma*
- Thermodynamics In Materials Science by *Robert T. Dehoff .*
- Fundamentals of Statistical and Thermal Physics, *F. Reif*
- Heat and Thermodynamics: A Historical Perspective (Greenwood Guides to Great Ideas in Science) by *Christopher J. T. Lewis.*
- Statistical mechanics, *R. P. Feynman, W. A. Benjamin.*
- Quantum Mechanics, *E. Merzbacher*
- DK Science Encyclopedia (Revised Edition) by Susan *McKeever and Martyn Foote.*
- The Cluster Expansion, *W. J. Mullin*
- Noise and Fluctuations, *D. K. C. MacDonald*
- Introduction to Metallurgical Thermodynamics by *David R. Gaskell.*
- An introduction to Statistical Thermodynamics, *T. L. Hill*
- A Guide to Physics: Thermodynamics, Statistical Physics, and Quantum Mechanics by Gerald D. Mahan, *Boris E. Nadgorny, and Max Dresden.*
- Equilibrium Statistical Physics, *M. Plische and B. Bergersen*
- Equilibrium Thermodynamics, *C. J. Adkins,*
- Course in Thermodynamics . Revised Printing. Volume

II. (Series in Thermal and Fluids Engineering) by *Joseph Kestin.*

- Thermodynamics (and Introduction to Thermostatistics), *H. B. Callen*
- Schaum's Outline of Thermodynamics for Engineers, 2nd edition (Schaum's Outlines) by *Merle Potter* and Ph.D., *Craig Somerton.*
- Introduction to Thermal Sciences: Thermodynamics Fluid Dynamics Heat Transfer by *Frank W. Schmidt*, *Robert E. Henderson*, and *Carl H. Wolgemuth.*
- Heat and Thermodynamics, *M. W. Zemansky*
- Mixing and Excess Thermodynamic Properties (Physical sciences data) by *Jaime Wisniak* and *Abraham Tamir.*
- A Modern Course in Statistical Physics, *L. E. Reichl*
- Schaum Engineering Thermodynamics (Schaum's Outlines) by *Merle Potter.*
- Essentials of Thermodynamics (Essentials) by Research & Education Association, Rea, and Staff of Research Education Association .
- Thermodynamics by *J.P. Holman.*
- Introductory Statistical Mechanics, *R. Bowley and M. Sánchez*
- The Language of Science by Sidney *B. Cahn.*